THE ORIGIN OF THE UNIVERSE

Other Books in the **SCIENCE MASTER SERIES**

THE ORIGIN OF THE
UNIVERSE

••

JOHN D. BARROW

BASIC
BOOKS

A Member of the
Perseus Books Group

Copyright © 1994 by John D. Barrow.
Published by BasicBooks,
A Member of the Perseus Books Group

• • • • •

• • • • •

Designed by Joan Greenfield

• • • • •

LIBRARY OF CONGRESS CATALOGING-IN-PUBLICATION DATA
Barrow, John D., 1952–
    The origin of the universe / John D. Barrow
        p.   cm. — (Science masters series)
    Includes bibliographical references and index.
    ISBN 0–465–05354–8 (cloth)
    ISBN 0–465–05314–9 (paper)
    1. Cosmology. 2. Astrophysics. I. Title. II. Series.
QB981.B2798 1994
523.1—dc20   94–6343
                                              CIP

9 8 7 6

To Dennis and Bill,

Cosmologists, gentlemen, and teachers,

To whom many owe much

....

Beautiful are the things we see

More beautiful those we understand

Much the most beautiful those we do not comprehend.

—Niels Steensen (Steno) 1638–1686

CONTENTS

We are living in the universe's prime, long after most of the exciting things have happened. Gaze into the sky on a starry night and you will see a few thousand stars, most straddling the darkness in a great swath we call the Milky Way. This is all the ancients knew of the universe. Gradually, as telescopes of greater and greater size and resolution have been developed, a universe of unimagined vastness has swum into view. A multitude of stars gathered into the islands of light we call galaxies, and all around the galaxies is a cool sea of microwaves—the echo of the big bang some fifteen billion years ago. Time, space, and matter appear to have their origins in an explosive event from which the present-day universe has emerged in a state of overall expansion, slowly cooling and continuously rarifying.

In the beginning, the universe was an inferno of radiation, too hot for any atoms to survive. In the first few minutes, it cooled enough for the nuclei of the lightest elements to form. Only millions of years later would the cosmos be cool enough for whole atoms to appear, followed soon by simple molecules, and after billions of years by the complex sequence of events that saw the condensation of material into stars and galaxies. Then, with the appearance of stable planetary environments, the complicated products of biochemistry were nurtured, by processes we still do not understand. But how and why did this elaborate sequence of events begin? What do modern cosmologists have to tell us about the beginning of the universe?

The various creation stories of ancient times were not scientific theories in any modern sense. They did not attempt to reveal anything new about the structure of the world; they aimed simply to remove the specter of the unknown from human imaginings. By defining their place within the hierarchy of creation, the ancients could relate the world to themselves and avoid the terrible consideration of the unknown or the unknowable. Modern scientific accounts need to achieve much more than this. They must be deep enough to tell us more about the universe than what we have put into them. And they must be broad enough to make predictions, as a check on their credentials to explain the things we already know about the world. They should bring coherence and unity to collections of disconnected facts.

The methods employed by modern cosmologists are simple, but not necessarily obvious to the outsider. They begin by assuming that the laws governing the workings of the world locally, here on Earth, apply throughout the universe until one is forced to conclude otherwise. Typically one finds that there are some places in the universe, especially in the past, where extreme conditions of density and temperature are encountered which are outside our direct experience on Earth. Sometimes our theories are expected to continue to work in these domains—and, indeed, do. But on other occasions we are working with approximations to the true laws of nature—approximations that possess known limits of applicability. When we reach those limits, we must try to establish better approximations to cover the unusual new conditions we have found. Many theories make predictions that we cannot test by observation. Indeed, it is those sorts of predictions that often dictate the types of observatory or satellite to be developed in the future.

Cosmologists often talk about constructing "cosmological models." By this they mean producing simplified mathematical descriptions of the structure and past history of the universe which capture its principal features. Just as a model airplane reproduces some, but not all, of the features of a real airplane, so a model universe cannot hope to incorporate every detail of the universe's structure. Our cosmological models are very rough and ready. They begin by treating the universe as if it were a completely uniform sea of material. The clumping of material into stars and galaxies is ignored. Only if one is investigating more specific issues, like the origins of stars and galaxies, are the deviations from perfect uniformity considered. This strategy works remarkably well. One of the most striking features of our universe is the way in which the visible part of it is so well described by this simple idealization of it as a uniform distribution of material.

Another important feature of our cosmological models is that they involve properties—like density or temperature—whose numerical values can be found only by observation, and only particular combinations of observed values for a number of these quantities will be allowed by the model. In this way compatibility between the model and the real universe can be checked.

Our exploration of the universe has taken off in different directions. Besides satellites, spacecraft, and telescopes, we have employed microscopes, atom-smashers and accelerators, computers and human thinking to enlarge our understanding of the entire cosmic environment. Besides the world of outer space—the stars, galaxies, and great cosmic structures—we have come to appreciate the labyrinthine subtlety within the depths of inner space. There we find the subatomic world of the nucleus and its parts: the basic building blocks of matter—so few in number, so simple in structure, but in combination capable of being organized into the vast panoply of complexity we see around us and of which we are a peculiar part.

These two frontiers of our understanding—the small world of the elementary parts of matter and the astronomical world of the stars and galaxies—have come together in unexpected ways in recent times. Where once they were the domains of different groups of scientists attempting to answer quite different questions by separate means, now their interests and methods are intimately entwined. The secret of how galaxies came into being may well be fathomed by the study of the most elementary particles of matter in particle detectors buried deep underground; the identity of those elementary particles may be revealed by observations of distant starlight. And as we try to reconstruct the history of the universe, searching for the fossil remnants of its youth and adolescence, we find that by the coming together of the largest and the smallest aspects of the physical world our appreciation of the unity of the universe becomes more impressive and complete.

This little book aims to provide a short account of the Beginning for beginners. What evidence do we have about the early history of the universe? What are the latest theories about how the universe could have begun? Can we test them by observation, and how does our own existence relate to them? These are some of the questions that will arise on our journey to the origins of time. I shall present some of the latest speculative theories about the nature of time, the "inflationary universe," and "wormholes," and along the way explain the significance of the COBE satellite observations that were greeted with such euphoria in the spring of 1992.

I would like to thank my cosmological colleagues and collaborators for their discussions and discoveries, which have made possible a modern story of the origin of the universe. Anthony Cheetham and John Brockman deserve credit for their conception of this project. It remains to be seen whether they were as wise to invite me to participate in it. I would also like to thank Gerry Lyons and Sara Lippincott for their editorial guidance. My wife, Elizabeth, has provided a vast amount of assistance, which helped bring things quickly to completion without pushing everything else off to infinity. As always, I am indebted to her for everything. Junior members of the family—David, Roger, and Louise—have seemed singularly unimpressed by the project. But they do like Sherlock Holmes.

Brighton
March 1994

THE ORIGIN OF THE UNIVERSE

....

CHAPTER I

THE UNIVERSE IN A NUTSHELL

"I must thank you," said Sherlock Holmes, "for
calling my attention to a case which certainly
presents some features of interest."
—*The Hound of the Baskervilles*

How, why, and when did the universe begin? How big is
it? What shape is it? What's it made of? These are questions that any curious child might ask, but they are also
questions that modern cosmologists have wrestled with for
many decades. One of the attractions of cosmology for
popular writers and journalists is that so many of the questions at the frontiers of the subject are easy to state. Look at
the frontiers of quantum electronics, DNA sequencing,
neurophysiology, or pure mathematics and you will not
find that the problems of the expert translate so readily
into the vernacular.

Until the early years of the twentieth century, neither
philosophers nor astronomers had questioned the notion
that space was absolutely fixed—an arena in which the
stars, the planets, and all the other heavenly bodies played
out their motions. But during the 1920s this simple picture
was transformed: first by the suggestions of physicists

exploring the consequences of Einstein's account of gravity, and then by the results of observations of light from stars in distant galaxies by the American astronomer Edwin Hubble.

Hubble made use of a simple property of waves. If their source moves away from the receiver, the frequency with which waves are received falls. To see this, wiggle your finger up and down in some still water and watch the wave crests moving off to some other point on the water's surface. Now move your finger away from that point as you make waves, and they will be received less frequently than they were emitted. Now move your finger toward the reception point, and the reception frequency goes up. This property is shared by all waves. In the case of sound waves, it is responsible for the change in pitch of a train whistle or a police siren as it passes you. Light is also a wave, and when its source is moving away from the observer the decrease in the frequency of the light waves means that visible light is observed to be slightly redder. Hence, this effect is called a "redshift." When the light source is approaching the observer, the reception frequency increases, visible light gets bluer, and it is called a "blueshift."

Hubble discovered that the light from the galaxies he was seeing displayed a systematic redshifting. By measuring the extent of the shift, he could determine how fast the sources of light were receding; and by comparing the apparent brightnesses of stars of the same sort (stars whose intrinsic brightnesses would be the same) he could deduce their relative distances away from us. What he discovered was that the farther away the source of light, the faster it was moving away from us. This trend is known as Hubble's Law, and its illustration with modern data is shown in figure 1.1. In figure 1.2 is shown an example of the light signal received from a

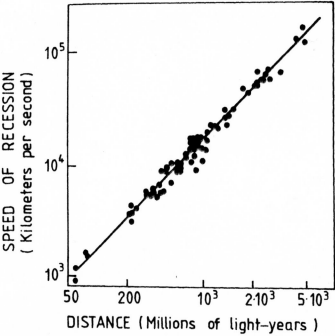

FIGURE I.I

A modern illustration of Hubble's Law, displaying the increase of recession speed of galaxies growing in direct proportion to their distance.

· · · · · · · · ·

distant galaxy, displaying the shift of the spectrum of various atoms toward the red, as compared with that emitted from the same atoms in the laboratory.

What Hubble had discovered was the expansion of the universe. Instead of a changeless arena in which we could follow the local perambulations of planets and stars, he found that the universe was in a dynamic state. This was the greatest discovery of twentieth-century science, and it confirmed what Einstein's general theory of relativity had predicted about the universe: that it cannot be static. The

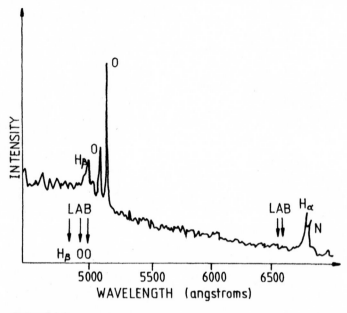

FIGURE I.2

The spectrum of a distant galaxy (known as Markarian 609), showing how three spectral lines (marked H_β, O, and O) near 5000 angstroms and two (marked H_α and N) near 6500 angstroms are systematically shifted toward higher wavelengths than they have when measured in the laboratory. The positions of the lines in the laboratory are indicated by the arrows marked LAB; the measured positions are the labeled peaks on the graph of the light spectrum. The shift toward the red (optical red light lies at about 8000 angstroms) enables the recession speed to be calculated.

· · · · · · · · ·

gravitational attraction between the galaxies would bring them all together if they were not rushing away from each other. The universe can't stand still.

If the universe is expanding, then when we reverse the direction of history and look into the past we should find evidence that it emerged from a smaller, denser state—a state that appears to have once had zero size. It is this apparent beginning that has become known as the big bang.

But we are going a little too fast. There are important things to appreciate about the present expansion of the universe before we start delving into the past. First of all, what exactly is expanding? In the movie *Annie Hall*, Woody Allen is found on his analyst's couch telling of his anxiety about the expansion of the universe: "Surely this means that Brooklyn is expanding, I'm expanding, you're expanding, we're all expanding." Thankfully, he was wrong. *We* are not expanding. Nor is Brooklyn. Nor is the Earth. Nor is the solar system. Nor, in fact, is the Milky Way galaxy. Nor even those aggregates of thousands of galaxies that we call "galaxy clusters." These collections of matter are all bound together by chemical and gravitational forces between their constituents—forces that are stronger than the force of the expansion.

It is only when we get beyond the scale of great clusters of hundreds and thousands of galaxies that we see the expansion winning out over the local pull of gravity. For example, our near neighbor the Andromeda galaxy is moving toward us, because the gravitational attraction between Andromeda and the Milky Way is larger than the effect of the universal expansion. It is the galaxy clusters, not the galaxies themselves, that act as the markers of the cosmic expansion. A simple picture might be to think of specks of dust on the surface of an inflating balloon. The balloon will expand and the dust specks will move apart, but the individual dust specks will not themselves expand in the same way. They act like markers of the amount of stretching of the rubber that has occurred. Similarly, it is best to think of the expansion of the universe as the expansion of the space between clusters of galaxies, as illustrated in figure 1.3.

FIGURE 1.3
The expansion of the universe viewed as the expansion of space. Mark points on the surface of a balloon to represent galaxy clusters and inflate it. The space between the clusters increases, but the size of the clusters does not. This is analogous to a universe with two dimensions of space, represented by the surface of the balloon. Any cluster on the inflating surface sees all the other clusters receding from it. Notice that the center of the expansion does not lie on the surface of the balloon.

. . . .

Next, we might worry about the implications of the fact that all the clusters are moving away from *us*. Why us? If we know anything about the history of science, it is that Copernicus demonstrated that the Earth is not at the center of the universe. Surely if we think that everything is moving away from *us* then we have reinstated ourselves back in the center of immensities. But this is not the case. The expanding universe is not like an explosion that has some origin at a point *in* space. There is no fixed background space into which the universe is expanding. The universe contains all the space there is!

Think of space as an elastic sheet. The presence and movement of material on this malleable space will produce indentations and curvature. The curved space of our universe is like the three-dimensional surface of a four-dimensional ball—something we cannot envisage. But imagine the universe as a flatland, with only two dimensions of space. It is then like the surface of a three-dimensional ball, which is easy to picture. Now imagine that this three-dimensional ball can get bigger—like our inflating balloon in figure 1.3. The surface of the balloon is an expanding two-dimensional universe. If we mark two points on it, those points will recede from each other as the balloon is inflated. Now put many marks all over the surface of the balloon and inflate it again. What you find is that at whatever mark you locate yourself, all the other marks will appear to expand away from *you* as the balloon expands. You will see a Hubble Law of expansion, with the widely separated marks receding from one another faster than the ones closer together. The lesson from this example is that the surface of the balloon represents space, but the "center" of the expansion of the balloon does not lie on that surface at all. There *is* no center of expansion on the balloon's surface. Nor is there any edge. You cannot fall off the edge of the universe; the universe is not expanding into anything. It is everything there is.

One question we might raise at this stage is whether the state of expansion we witness in the universe will continue indefinitely. If we throw a stone in the air, it will return to Earth, pulled back by the force of the Earth's gravity. The harder we throw, the more energy we give the moving stone, and the higher it will go before it returns. Now, we know that if we launch a projectile faster than 11 kilometers per second it will escape the pull of the Earth's gravity. This is the critical launch speed for rockets. Space scientists call it the "escape velocity" of the Earth.

Similar considerations apply to any exploding or expanding system of material retarded by the pull of gravity. If the energy of outward motion exceeds that created by the inward pull of gravitation, the material will exceed the escape velocity and just keep on expanding. But if the attractive pull that gravity exerts between its parts is the greater, the objects in the expansion will eventually start to come back together again, just as the Earth and the stone do. So it is with expanding universes; there is a critical launch speed at the start of their expansion. If the speed exceeds this, the gravitational pull of all the material in such a universe will not be able to halt the expansion, and it will keep expanding forever. On the other hand, if the launch speed is less than the critical value, eventually the expansion will halt and reverse, culminating in a contraction back to zero size—the very same state in which it apparently began. In between, there exists what I call the "British compromise universe," which has exactly the critical launch speed—that is, the smallest value that will keep it expanding forever (see figure 1.4). One of the great mysteries about our universe is that it is currently expanding tantalizingly close to this critical case. So close, in fact, that we cannot yet say for sure on which side of the critical divide it lies. We do not know what the long-range forecast is.

Cosmologists regard the fact that we are so close to this critical divide as a peculiar property of the universe which requires an explanation. It is difficult to understand because, as the universe expands and ages, it will diverge farther and farther from the critical divide if it does not begin with precisely the critical launch speed. This creates a major puzzle. The universe has been expanding for about fifteen billion years, yet it is still so close to the critical divide that we cannot tell on which side it lies. To have remained so close after such a huge period of time turns out to require the universe's launch speed to have been "chosen" to differ from the critical one by no more than one part in ten followed by thirty-five zeros. Why?! Later, we shall see that our study of what might have happened during the first moments of the universal expansion offers a possible explanation for this highly unlikely state of affairs. But for now we shall content ourselves with under-

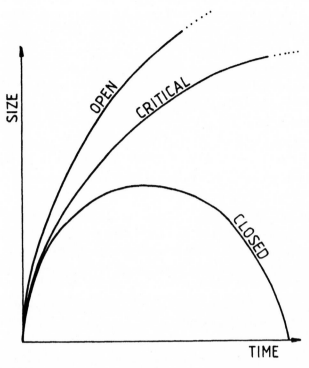

FIGURE I.4

The three varieties of expanding universe. "Open" universes are infinite in extent and expand forever. "Closed" universes are finite and contract back to a "big crunch." The divide between the two is marked by the "critical" universe, which is infinitely large and expands forever.

··········

standing why any universe that contains human beings has to lie very close to that critical divide after billions of years of expansion.

If the universe starts expanding far faster than the critical speed, then gravity can never draw together local islands of material to form galaxies and stars. The formation of stars is a crucial step in the evolution of a universe that is to be observed. Stars are condensations of matter large enough to create at their centers pressures great enough to initiate spontaneous nuclear reactions. Those reactions burn hydrogen into helium throughout a long and sedate period of their history—a period that our sun is in the midst of—but in the final stages of their lives stars encounter a nuclear energy crisis. They undergo an explosive period of rapid change, in which helium is transformed into carbon, nitrogen, oxygen, silicon, phosphorus, and all the other elements that play a vital role in biochemistry. When stars explode in supernovas, these elements are dispersed into space and ultimately find their way into planets and people. The stars are the source of all the elements upon which complexity, and therefore life, is based. Every nucleus of carbon in our bodies originated in the stars.

So we see that universes which expand faster than the critical divide will never give birth to stars, and hence will never produce the building blocks required to make "living" entities as complex as human beings or silicon-based computers. Similarly, if a universe expands at far less than the critical speed, its expansion will be reversed into contraction before the stars have had time to form, explode, and create the constituents of living things. Again, we are left with a universe unable to give rise to life.

Thus we learn a surprising lesson: only those universes that still expand very close to the critical divide after billions of years can produce the material out of which any structure complex enough to qualify as an observer must be made (see figure 1.5). We should not be surprised to find our universe expanding so close to the critical divide. We could not exist in any other sort of universe.

The development of our picture of the expanding universe and the reconstruction of its past history moved very slowly.

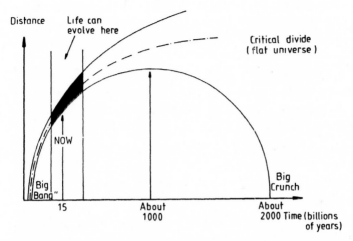

FIGURE 1.5

Universes that are too far above the critical divide expand too fast for matter to condense into stars and galaxies; such universes therefore remain devoid of life. Those that fall too far below the critical divide collapse before stars form. The shaded region indicates the range of cosmological expansions and epochs in which observers could evolve.

• • • • • • • • •

During the 1930s, the Belgian priest and physicist Georges Lemaître played a leading role in its inception. His theory of the "primeval atom" was a precursor of what is now known as the big-bang model. The most important steps were taken during the late 1940s by George Gamow, a Russian émigré to the United States, together with two of his young research students, Ralph Alpher and Robert Herman. They began to take seriously the possibility of applying known physics to figure out what the early stages of an expanding universe might have been like. They recognized one key point: if the universe began in a hot, dense state in the distant past, there should remain some radiation from this explosive beginning. More specifically, they realized that when the universe was just a few minutes old it should have been hot enough for nuclear reactions to occur everywhere. Later, these important insights would be confirmed by much more detailed predictions and observations.

In 1948, Alpher and Herman predicted that the remnant radiation from the big bang, having been cooled by the expansion of the universe, should now have a temperature of around five degrees above absolute zero (absolute zero is equal to -273°C)—that is, five degrees Kelvin. Their prediction lay buried in the physics literature, however. A decade and a half later, several other scientists were considering the problem of the origin of a hot, expanding universe, but none of them knew about Alpher and Herman's paper. Communications were not then what they are now. Reconstructing the details of the universe's early history was not a very serious activity in the minds of most physicists in the 1950s and early 1960s. But in 1965, everything changed. Alpher and Herman's cosmic radiation field—manifested as microwave noise coming with the same intensity from all directions in the sky—was discovered serendipitously by Arno Penzias and Robert Wilson, two radio engineers at Bell Labs, in New Jersey, who were calibrating a sensitive radio antenna for tracking the first Echo satellite. Meanwhile, only a few miles away at Princeton University, a group led by the physicist Robert Dicke had independently recalculated what Alpher and Herman had long ago published, and had set about designing a detector to mount a search for remnant radiation from the big bang. They learned of the unexplained noise in the Bell Labs receiver and soon interpreted it as the relic radiation they were looking for. If the source was indeed heat radiation, the temperature was 2.7°K—very close to Alpher and Herman's inspired estimate. The phenomenon was dubbed the "cosmic microwave background radiation."

The discovery of the cosmic microwave background marked

the beginning of the serious study of the big-bang model. Gradually, other observations revealed further properties of the background radiation. It had the same intensity in every direction to at least one part in a thousand. And as its intensity was measured at different frequencies, it began to reveal the characteristic variation of intensity with frequency which is the signature of pure heat. Such radiation is called "blackbody" radiation. Unfortunately, the absorption and emission of radiation by molecules in the Earth's atmosphere prevented astronomers from confirming that the whole spectrum of the radiation was indeed that of heat radiation. Suspicions remained that it might have been produced by violent events that occurred nearby in the universe long after the expansion began. These doubts could be overcome only by observing the radiation from above the Earth's atmosphere, and measuring the whole spectrum from space was the first great success of NASA's Cosmic Background Explorer (COBE) satellite in 1989. It was the most perfect blackbody spectrum ever seen in nature, and a striking confirmation that the universe was once hundreds of thousands of degrees hotter than it is today (see figure 1.6). For only under such extreme conditions could the radiation in the universe assume a blackbody form to such high precision.

Another key experiment to confirm that the background radiation did not have a recent origin nearby in the universe was carried out by high-flying U2 aircraft. These former spyplanes are extremely small, with large wingspans, which makes them very stable platforms for making observations. On this occasion, they were looking up rather than down, and they detected a small but systematic variation in the intensity of the radiation around the sky—a variation which had been predicted to appear if the radiation had originated in the distant past. If the radiation formed a uniformly expanding sea, emerging from the early stages of the universe, then we would

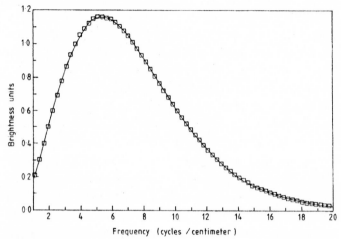

FIGURE 1.6

The variation of the intensity of the microwave background radiation with its frequency, as observed by the COBE satellite from above the Earth's atmosphere. The observations (boxes) display a perfect fit with the (solid) curve expected from pure heat radiation with a temperature of 2.73°K.

· · · · · · · · · ·

be moving through it. The aggregate of the Earth's motion around the sun, the sun's motion around the Milky Way, the Milky Way's motion among its neighbors, and so on, means that we are moving through the radiation in some direction (see figure 1.7). The radiation intensity will appear greatest when we look in that direction and least intense 180 degrees away, and should display a characteristic cosine variation with angle in between (see figure 1.8). It is rather like running in a rainstorm. You get wettest on your chest and stay driest on your back. Here it is microwaves that are swept up in our net direction of motion. The observations revealed a perfect cosine variation, as predicted.

Subsequently, several different experiments confirmed the discovery of "The Great Cosine in the Sky," as it became known. We, and the local cluster of galaxies in which we reside, are moving relative to the sea of cosmic microwaves. The radiation cannot therefore have arisen locally, because it would then have shared our motion, and the cosine variation in its intensity would not have been seen.

Our motion through the background radiation from the big bang is not the only thing that can cause its intensity to vary slightly from one direction to another. If the universe is expanding at slightly different rates in different directions, the radiation will be less intense (cooler) in the directions of faster expansion. Moreover, there are large concentrations of matter, as well as regions devoid of matter, in some directions; these, too, should alter the intensity of radiation coming from those directions. It was the search for these variations that motivated the COBE satellite mission, and their discovery that made headline news all over the world in 1992.

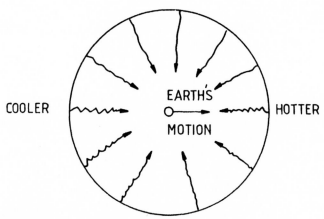

FIGURE 1.7

Our motion through the isotropic cosmic sea of microwaves arriving from the big bang. We measure the maximum intensity in the direction we are moving, and a minimum in the opposite direction, with a steady cosine variation in between.

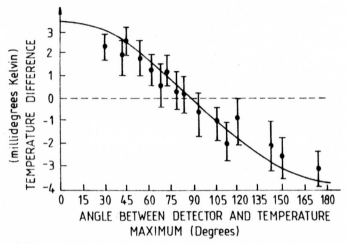

FIGURE 1.8

"The Great Cosine in the Sky" shows the actual differences in temperature of the microwave background radiation in millidegrees Kelvin, as one varies the angle of observation from the direction in which it is maximum to the direction in which it is minimum. The error bars show the accuracy of each temperature measurement.

· · · · · · · · ·

When we examine all these measurements of the intensity of radiation coming to us from different directions in the sky, we learn a number of striking things about the structure of the universe. We find that it is expanding at the same rate in every direction to an accuracy better than one part in a thousand. We say that the expansion is isotropic—that is, the same in every direction. If one had been picking possible universes at random from some cosmic menagerie, there would be countless varieties that expand far faster in some directions than others, or that rotate at high speed, or even that contract in some directions while expanding in others. Yet our universe is peculiar: it seems to be in an improbably well-ordered state, in which the expansion proceeds at the same rate in every direction to high precision. It is as if you were to find all your children's bedrooms perfectly tidy—a highly unlikely state of affairs. You decide that some outside influence must have been exerted. Likewise, there must be some explanation for the striking isotropy of the expansion.

Cosmologists have long regarded the isotropy of the universe's expansion as a great mystery that must be explained. The approaches taken to it illustrate something of the styles of thinking within the subject. The first line that could be taken is to say that the universe began expanding isotropically from the outset and that the present state is just a reflection of its special starting conditions. Things are as they are because they were as they were. As it stands, this is not very helpful. It doesn't explain anything. It's like invoking the tooth fairy. But it could, of course, be true. If so, we might hope to find some deep "principle" calling for an initial state of isotropic expansion. Such a principle might have other, more local applications by which it could reveal itself. The unsavory feature of this approach is that it places the onus for explaining the present state of the universe entirely upon its unknown (and perhaps unknowable) starting state.

The second approach is to regard the present state of affairs as a consequence of physical processes still going on in the universe. Thus perhaps no matter how irregular was its initial state, after billions of years the irregularities all get washed out, leaving a state of isotropic expansion. This approach has the merit of suggesting possible research programs: Are there cosmic processes that can smooth out nonuniformities in the expansion? How long does the smoothing take? Can these processes get rid of any amount of irregularity by the present day, or can they eradicate only a small amount? This approach allows us to say that no matter how the universe began, there are processes inevitably arising within it during its early history which insure that after fifteen billion years of expansion it looks pretty much the way it does today.

Although the second philosophy sounds wonderfully appealing, it does have a downside. If we succeed in showing that the present state of the universe emerges regardless of the starting conditions, then our observations of its structure will not be able to tell us anything about those starting conditions—for the present state would be compatible with any starting state. But if, on the contrary, the present structure of the universe—its expansion isotropy, and the patterns displayed by the clustering of galaxies—are partial reflections of the way the universe began, then it might be possible to determine something about the initial state of the universe by observing it today.

··

CHAPTER 2

THE GREAT UNIVERSAL CATALOG

All other men are specialists, but his specialism is
omniscience.
—*The Bruce-Partington Plans*

When Einstein published his general theory of relativity in
1915, there was no widespread belief that the universe was
populated by those huge collections of stars we know as
galaxies. It was commonly held that these extraterrestrial
sources of light—or "nebulae," as they were then called—
lay within our own Milky Way galaxy. Nor had there ever
been any proposal by astronomers or philosophers that the
starry universe was anything but static. It was into this
intellectual ambience that Einstein launched his new
theory of gravitation. Unlike Newton's classical description
of gravitational forces, which Einstein's theory included and
superseded, the general theory of relativity had the extraordinary ability to describe entire universes, even if they were
infinite in extent. Only the simplest of solutions to Einstein's
equations have ever been found. Fortunately, the very simple
ones describe rather well the universe we see.

When Einstein began to explore what his new equations
revealed about the universe, he set about doing what scien-

tists generally do—simplifying the problem to be solved. The real universe, with all its bells and whistles, was much too complicated a beast to deal with, so he simplified it by assuming that matter is uniformly distributed everywhere. That is, he ignored the variations in the density of matter from place to place which constitute the heavenly bodies. He also assumed that the universe looked the same in every direction. These, we now know, are excellent approximations of the state of our universe, and cosmologists still make them today when they want to deduce things about the overall evolution of the universe. But what Einstein then discovered to his great chagrin was that his equations demanded that universes of this sort either expand or contract with the passage of time. There is no great mystery about this. It is true even in Newton's description of gravity. If you place a cloud of dust particles in space, they will begin to feel a mutual gravitational attraction; the cloud will gradually contract. The only circumstance that will prevent this is some sort of explosion that drives the particles apart. They cannot remain in an unchanging state unless another force intervenes to oppose gravity. In the absence of that opposing force, the gravitational attraction between a static distribution of stars and galaxies will cause them to fall in upon themselves.

Einstein was deeply troubled by this prediction of his theory. Apparently, he lacked the confidence to claim that the universe is not static. An expanding universe was a very strange notion at the time. Instead, he set about exploring ways to legitimately modify his new theory of gravity so as to suppress the possibility for expansion or contraction of the universe. He noticed that the mathematics allowed a term to be introduced representing a force of repulsion that opposed the pull of gravity on pieces of matter. If he included this term—which he called the "cosmological constant"—in his general theory of relativity, Einstein could now find a model in which the repulsion exactly counterbalanced the attraction of gravity. This model has become known as the Einstein static universe (see figure 2.1).

In 1922, Alexander Friedmann, a young mathematician and atmospheric physicist in St. Petersburg, studied Einstein's calculations and became convinced that the master had made a crucial oversight. The static universe was certainly *a* solution of the modified equations, but it was not the only one. There were others, which described the expanding universes that the original equations had demanded. The expansion of the real universe could not be avoided by Einstein's antigravitational force. Friedmann found all the possible expanding universes that the general-relativity equations allowed, and transmitted his results to Einstein. At first, Einstein thought Friedmann had simply calculated wrongly.

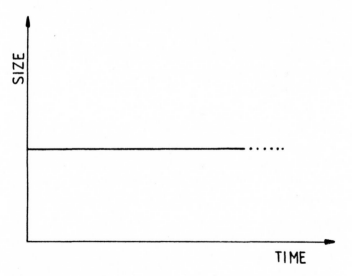

FIGURE 2.1
A static universe has a size that does not change with time. It has no beginning and no end.

But he was soon persuaded otherwise by Friedmann's colleagues, and he realized that the inclusion of the cosmological constant produced an unrealistic static universe: if Einstein's static universe were altered in the tiniest degree, it would begin to expand or contract. It was the cosmic version of a needle balanced on its point.

Many years later, Einstein referred to his espousal of the cosmological constant as "the greatest blunder of my life." By introducing it into his equations, he had missed the opportunity to make the sensational prediction that our universe is expanding. That distinction fell to Alexander Friedmann. Sadly, Friedmann did not live to see the confirmation of his prediction by Edwin Hubble's observations seven years later and the eventual acceptance of the paradigm of the expanding universe. Friedmann's meteorological research led him to make many dangerous high-altitude balloon flights—he held the world altitude record for a time—and in 1925 he died from the aftereffects of one such flight. His death was a great loss to science. He was just thirty-seven years old.

Although Einstein had inherited the traditional notion of a static universe, this does not mean that his predecessors denied the possibility that any changes could occur to the state of the universe. Although there was no previous conception of universal expansion or contraction, there had been much speculation that the universe might be winding down into an increasingly disordered and uninhabitable state. This expectation sprang from the study of how heat can be used as a source of power. The Industrial Revolution had led to a number of advances in science and engineering, the most important of which was the design and understanding of machines and steam engines. Out of these developments grew the study of heat as a form of energy. It was understood that energy was a conserved commodity. It could neither be created nor destroyed, merely changed from one form into another.

But there was more to it than that. Some forms of energy are more useful than others. The measure of their usefulness is a measure of the order of the form in which the energy exists: the more disordered, the less useful. This disorder, which became known as "entropy," always seems to increase in natural processes. At one level, there is no mystery about this. Your desk top and your children's bedrooms seem to evolve from a state of order into one of disorder—but never the reverse. There are just so many more ways for things to go from order to disorder than vice versa that the former tendency is the one we see in practice. This idea became enshrined in the famous "second law of thermodynamics," which states that the entropy of a closed system never decreases.

The fascination with heat-driven engines led Rudolf Clausius—the formulator of the second law in 1850 and the inventor of the term "entropy"—and others to consider the universe itself as a closed system, subject to the same thermodynamic laws. This created a somewhat pessimistic long-range outlook: everything seemed to be heading for an uninteresting, structureless state, to which all the ordered forms of energy in the universe would ultimately find themselves degraded. Pursuing these ideas to their logical conclusion, Clausius introduced the concept of the "heat death" of the universe; he predicted that in the future "the universe would be in a state of unchanging death," because the entropy would steadily increase until it had attained its maximum possible value, after which no changes could occur. The universe would be left in its maximum-entropy state—a featureless sea of radiation, everywhere the same. There would be no ordered things like stars, or planets, or life—just heat radiation, getting cooler and cooler until a final equilibrium is reached.

Others began to examine the consequences of this idea for the dim and distant past. It seemed to imply that the universe must have had a beginning—a state of maximum order. In 1873, an influential British philosopher of science, William Jevons, claimed:

> We cannot . . . trace the heat-history of the Universe to an infinite distance in the past. For a certain negative [that is, past] value of the time the formulae give impossible values, indicating that there was some initial distribution of heat which could not have resulted, according to known laws of Nature, from any previous distribution. . . . Now the theory of heat places us in the dilemma either of believing in Creation at an assignable date in the past, or else of supposing that some inexplicable change in the working of natural laws then took place.

Strikingly, this argument for a beginning to the universe was presented fifty years before the notion of the expanding universe. It was reiterated by the British astrophysicist Arthur Eddington during the 1930s, in the context of the expanding universes that emerged from Einstein's theory of gravitation and confirmation of the expansion by Hubble's observations. He wrote:

> Following time backwards we find more and more organisation in the world. If we are not stopped earlier, we go back to a time when the matter and energy of the world had the maximum possible organisation. To go back further is impossible. We have come to another end of space-time—an abrupt end—only according to our orientation we call it "the beginning.". . . I find no difficulty in accepting the consequences of the present scientific theory as regards the future—the heat death of the universe. It may be billions of years hence, but slowly and inexorably the sands are running out. I feel no instinctive shrinking from this conclusion. . . . It is curious that the doctrine of the running down of the

physical universe is so often looked upon as pessimistic and contrary to the aspiration of religion. Since when has the teaching that "heaven and earth shall pass away" become ecclesiastically unorthodox?

The "heat death of the universe" became increasingly well known during the 1930s, due to its popularization in the widely read books by Eddington and his countryman the astrophysicist James Jeans. The wedding of the picture of an ever-expanding universe to Clausius's heat death only exacerbated the notion of a steady dissolution of the universe's contents into structureless heat radiation. One can find the pessimism this notion inspired permeating many theological and philosophical writings of those times, emerging even in the works of such contemporary novelists as Dorothy Sayers. It signaled the inevitable extinction of life not only on Earth but elsewhere as well, confirming the message of the man with the sandwich board that the end of the world was, if not nigh, at least on its way.

It is interesting to note that the arguments for a beginning made by Jevons and others are not quite correct, yet nobody seems to have noticed why at the time. Although the second law of thermodynamics requires the entropy of the universe to get smaller as we trace it back into the past, this does not mean that it need ever reach zero after a finite time, as it does in figure 2.2. The entropy could increase exponentially with time and just continue getting closer and closer to zero in the past without ever actually reaching zero, as in figure 2.3.

Alternatively, the entropy of the universe could increase with the passage of time, while that in a local region decreased. This is what occurs in many places at present. As the Earth's biosphere becomes more ordered locally, its entropy decrease is more than paid for by the overall entropy increase that occurs when the heat exchange

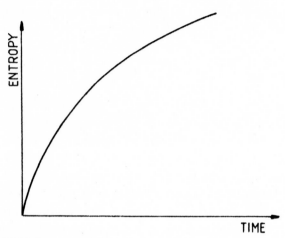

FIGURE 2.2
An increase of entropy from a state where the entropy was zero at a finite time in the past.

· · · · · · · · · ·

between the Earth and the sun is included in the accounting. If you set about making a chair out of some pieces of wood, then the level of order goes up in the construction process—entropy decreases. Yet there is no violation of the second law of thermodynamics, because the total entropy—which includes the expenditure of the energy stored in starches and sugars in our bodies and used in the work we did—increases. Indeed, the complexity of the living world we see around us is a manifestation of the subtle ways in which nature can create local decreases in entropy which are more than balanced by increases elsewhere.

It is only recently that cosmologists have realized that the predicted heat death of ever-expanding universes in a future state of maximum entropy will not occur. Although the entropy of the universe will continue to increase, the maximum entropy it can have at any given time increases even faster. Thus the gap between the maximum possible entropy and the true entropy of our universe continually widens, as shown in figure 2.4. The universe actually gets farther and farther away from the "dead" state of complete thermal equilibrium.

When we calculate the present entropy of the universe, we find that it is staggeringly low; that is, we can conceive of distributing the forms of energy in the universe in ways that are far more disordered. The universe is still in a highly ordered state, despite having expanded in an entropy-increasing manner for fifteen billion years. This is a puzzle. It implies that the starting state of the universe must have been very highly ordered, and hence extremely special and perhaps governed by some grand principle of symmetry or economy. However, it has proved impossible to exploit these ideas to discover that principle, because we do not know enough about the universe's structure to identify all the ways in which order and disorder are present

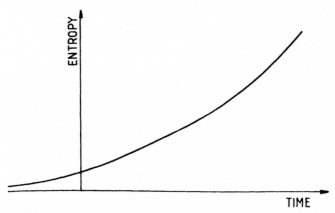

FIGURE 2.3
Another possible universe, in which the entropy is always increasing but gets closer and closer to zero in the past without ever reaching zero.

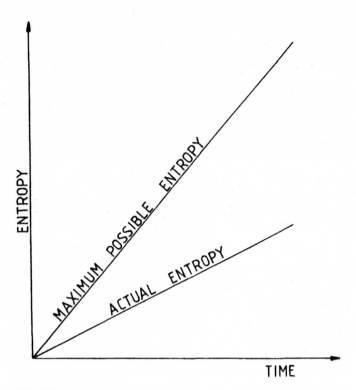

FIGURE 2.4

The modern perspective on the "heat death" of the universe. The actual entropy of a universe that expands forever increases continuously with time, but the maximum possible entropy of a universe containing the same amount of matter increases more rapidly. With the passage of time, the universe will therefore get farther and farther away from a "heat death" of complete equilibrium at its maximum possible entropy.

· · · · · · · · · ·

within it. This makes our calculation of its present entropy incomplete. For example, in 1975 the physicists Jacob Bekenstein and Stephen Hawking showed that black holes possess an entropy associated with their deep quantum aspects. The British mathematician Roger Penrose has speculated that an analogous entropy might also be associated with the gravitational field of the universe. A full understanding of the thermodynamic aspects of gravity is a problem for the cosmologists of the future. We shall return to it at the very end of our story.

If you don't fancy an ever-expanding universe heading for a lifeless future of ever-increasing entropy, you could pick another of Alexander Friedmann's expanding-universe models. Some expand slowly enough for the gravitational pull of matter to cause them to contract back to zero size in the far future. Their final state would be a heat death with a vengeance, with temperatures and densities increasing without limit as the contraction intensified. This model of cosmic evolution suggests the ancient idea of the cyclic universe—one that undergoes a never-ending sequence of rebirths, each time rising phoenixlike from the ashes of its previous demise (see figure 2.5).

According to this view, we are living in one expand-

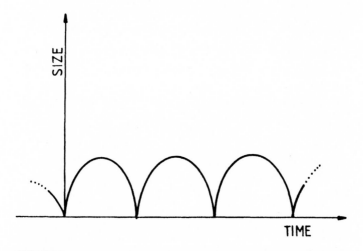

FIGURE 2.5
A possible eternally oscillating universe in which each cycle is the same size as its predecessor.

ing cycle of an infinitely old oscillating universe with an infinite future. All the planets, stars, and galaxies would be destroyed each time the universe plunged down to a "big crunch" and bounced back into a state of expansion. While this is philosophically appealing to some—and does away with the need to explain what happened at the beginning of the universe to produce the present state of expansion—it is also subject to criticism, because of the second law of thermodynamics. This argument was introduced in the 1930s by the American physicist Richard Tolman, who noted that the size of the universe would increase at each maximum and therefore each cycle would be larger than the last one. This would occur because the gradual dissipation of matter into radiation would increase the pressure opposing gravity, and the expansion would thus last longer in each ensuing cycle. So if we follow an oscillating universe backward in time, it gets smaller and smaller. Again, at the time (and for long afterward), it was wrongly concluded that this implies that the universe started expanding from zero size at a finite time in the past. It may have, but equally there could have been an infinite number of past cycles, each larger than its predecessor, without zero size ever being reached (see figure 2.6).

Others argued that if there had been an infinite number of oscillations in the past, then the increase of entropy would have led to a heat death by the present. However, since no one could be sure what went on at each bounce, this was not a very persuasive argument. Some people speculated that the constants of physics, the entropy, or even all of nature's laws might be redealt at each bounce. One finds little weight placed upon this argument today, because we lack a full understanding of what contributes

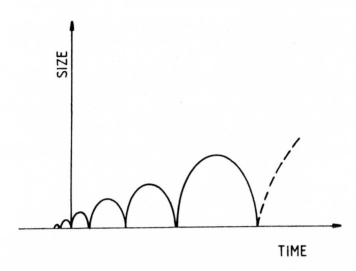

SIZE

TIME

FIGURE 2.6

The continuous increase of entropy with time, in accord with the second law of thermodynamics, increases the pressure of radiation in the universe and makes the cycles increase in size as time passes.

•••••••••

to the entropy of the universe. If the gravitational field carries entropy in unusual ways, a continual increase in the universe's entropy may well not result in a steady growth in its size from cycle to cycle.

Talk to anyone who is not an astronomer but who has a passing interest in the subject and you may find that mention of the big-bang theory provokes a recollection of something called "the steady-state theory of the universe." In fact, the steady-state theory ceased to be of interest to cosmologists about thirty years ago, but it lives on in the popular mind nonetheless, as a rival to the big bang. It was the brainchild of the astrophysicists Thomas Gold, Hermann Bondi, and Fred Hoyle, who thought it up at Cambridge University in 1948, after an excursion to see *The Dead of*

Night, a film that ends by returning to the circumstances in which it began. What if the universe were like that? they asked themselves. They knew that the universe was expanding but disliked the idea that it should have a beginning, which expansion seemed to imply. They wanted the universe to present the same general appearance to observers at all times from an infinite past to a future eternity. So they conceived a model in which the universe was, on the average, always the same and had no beginning (see figure 2.7).

They proposed that material, instead of having been created at one special moment in the past, was *always* being created, and at just the right rate to balance the dilution caused by the expansion, thus maintaining a constant density of material in the universe. This state of affairs had existed from a past eternity and would continue forever. By contrast, the big-bang picture of the expanding universe has a decreasing density, an apparent beginning, and no ongoing creation. Incidentally, the creation rate that the

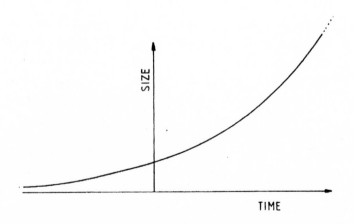

FIGURE 2.7
The expansion of a steady-state universe. It has no beginning and no end.

steady-state picture required is amazingly small (about one atom in a cubic meter every ten billion years), and there is no possibility of ever observing such a slow creation process directly. The reason that it is so small is that there is very little material in the universe. If all the stars and galaxies in the universe today were smoothed out into a uniform sea of atoms, there would only be about one atom in every cubic meter of space. This is a far better vacuum than could ever be produced in a laboratory on Earth. Outer space really is mostly that—space.

One of the merits of the steady-state proposal is its definiteness. It makes very strong predictions about what the universe should be like, and therefore is susceptible to being disproved by observations. And so it was. If the universe is to look the same at all cosmic epochs, there must not be particular periods of cosmic history when special things happen—when, for example, galaxies start to form, or when quasars are prevalent. The new science of radio astronomy was emerging from the wartime studies of radar. It enabled astronomers to look at objects that emitted their energy predominantly in the form of radio waves rather than as visible light. Astronomers used radio telescopes to observe very old galaxies that were strong sources of radio waves, to see whether galaxies of this sort appeared in the universe at a particular epoch, as the big-bang theory would predict, or were always equally abundant, as the steady-state theory predicted. During the late 1950s, observations started to accumulate indicating that the universe was very different in the past from what it is today. The galaxies that were strong sources of radio waves were not equally abundant at all times in cosmic history.

When we observe the light from distant astronomical objects, we are seeing them as they were in the past, at the time when the light left them, and so our observations of intrinsically similar objects at different distances allow us to examine what the universe was like at different times. It was, of course, still possible to dispute what these observations were telling us, and a vigorous debate arose as radio astronomers tried to convince the steady-state advocates that radio galaxies were much more numerous in the distant past than they are today. It was at this time that the big-bang versus steady-state argument impressed itself on the popular mind. The scene had already been set by a particularly influential series of BBC radio talks entitled "The Nature of the Universe," delivered by Fred Hoyle in 1950, in which he coined the term "big bang" as a pejorative description of a cosmology in which the universe expanded into being from a dense state at a finite time in the past.

This complicated dispute was finally settled in 1965, when the microwave background radiation was discovered by Penzias and Wilson. No such heat radiation would be present in a steady-state universe, because such a universe would not have experienced a hot past of enormous density; rather, it would always have been, on the average, cool and quiescent. Moreover, subsequent observations of the abundances of the lightest elements in the universe matched the predictions of the big-bang model and confirmed the idea that they were produced by nuclear reactions during the first three minutes of the expansion. The steady-state model offers no natural explanation for these abundances, because it never experiences an early period of great density and temperature when nuclear reactions can occur throughout the universe.

These two successes were the death knell for the steady-state model, and it played no further role as a viable model for the universe, despite attempts by some of its advocates to modify it in various ways. The big-bang model has established itself as the one that succeeds in coordinating our observations of the universe. But one must understand that the term "big-bang model" has come to mean nothing more than a picture of an expanding universe in which the past was hotter and denser than the present. There are many different cosmologies of this general type. The job of cosmologists is to pin down the expansion history of the universe—to determine how the galaxies formed; why they cluster as they do; why the expansion proceeds at the rate that it does—and to explain the shape of the universe and the balance of matter and radiation existing within it.

....

CHAPTER 3

THE SINGULARITY AND OTHER PROBLEMS

Singularity is almost invariably a clue. The more featureless and commonplace a crime is, the more difficult it is to bring home.

—*The Boscombe Valley Mystery*

The picture of an expanding universe implies that something cataclysmic must have occurred in the past. If we reverse the expansion of the universe and trace it backward in time, we appear to encounter a "beginning," at which everything hits everything else: all the mass in the universe is compressed into a state of infinite density. This state is known as the "initial singularity." The specter of its presence in our past has sparked all manner of metaphysical and theological extrapolations of the ideas of modern cosmology.

Judging by the rate at which the universe is observed to be expanding today, and the rate at which that expansion is decelerating, the initial singularity lies only about fifteen billion years ago. I say "only" because this time scale, although stupendously long by human standards, is not that much greater than more parochial time scales: dinosaurs walked in Argentina two hundred and thirty million years ago; the oldest fossilized bacteria found on Earth are about three billion years old; the oldest surface rocks in the Greenland mantle are 3.9 billion years old, and the oldest debris left over from the beginnings of our solar system are about 4.6 billion years old. The period of time that separates us from the origin of the Earth is barely one third of that which separates us from the mysteries of the singularity.

In the early 1930s, many cosmologists were loath to believe that the expansion really pointed to a singular beginning of infinite density. Two objections were raised. If we try to squeeze a balloon down to smaller and smaller size, we find our efforts opposed and ultimately defeated by the pressure exerted by the molecules of air within the balloon. As the volume in which they are free to move is decreased, they beat harder upon its boundaries. Likewise with the universe; we would expect the pressure exerted by the matter and radiation within the universe to prevent its ever being squeezed to zero volume. It might rebound, like a collection of colliding pool balls. Others claimed that the idea of an initial singular point of infinite density arose only because we had adopted a picture in which the universe was expanding at the same rate in every direction. Thus when the expansion was traced backward everything arrived at one point simultaneously. If, however, the expansion were slightly asymmetrical (and in reality it is), then when we traced it backward the imploding material would be out of step, so it might well avoid producing a singularity.

When these objections were explored, they failed to remove the expected singularity. In fact, the addition of pressure actually assisted its creation, because of Einstein's famous discovery that energy and mass are equivalent ($E=mc^2$). Pressure is just another form of energy, and thus is equivalent to mass; when it grows very large, it creates a gravitational force that opposes the repelling effect we usually associate with a pressure. Trying to avoid the singularity by increasing the resisting pressure was self-defeating; it actually made the singularity worse! Moreover, when Einstein's theory of gravitation was used to find other possible types of universes—universes that expand at different rates in different directions, or possess variations from place to place—the singularity remained. It was not just an artifact of symmetrical universe models. It seemed to be ubiquitous.

The last objection raised to the conception of the initial singularity was more subtle, and not fully understood until 1965. It is best illustrated by a more familiar situation. On a geographer's globe of the Earth, one finds a network of lines of latitude and longitude, which are used to specify the position of any point on the Earth's surface. As we move toward either of the poles, the lines of longitude begin to converge, and the meridians eventually intersect at the poles (see figure 3.1). So we see that at the poles the map coordinates have developed "singularities," although no real peculiarity has arisen on the Earth's surface. We have created an artificial singularity by a particular choice of map coordinates. We can always pick a different grid of coordinates, in which nothing bad happens at the poles. How do we know that the apparent singularity at the start of an expanding universe is not just an artifact of a poor way of mapping what has happened in the distant past?

To deal with these objections cosmologists had to be careful how they defined a singularity. If we envisage the entire history of the universe—all of space and all of time—like a vast sheet stretching out in front of us, singularities where the density and temperature becomes infinitely great might be found at particular places. Now suppose we cut around these pathological points and remove them to create a perforated sheet that contains no

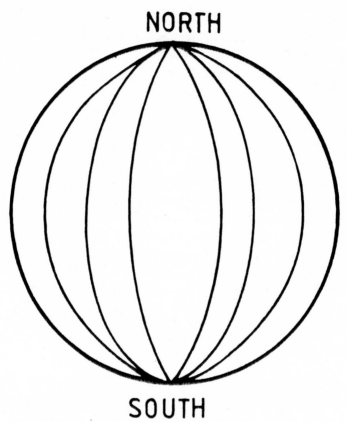

NORTH

SOUTH

FIGURE 3.1

Meridians mapping the Earth's surface intersect at the poles.

....

singularities. This would be another possible universe. But we feel cheated by this gambit. Surely such a universe is "almost" singular in some sense. And if we ever found a nonsingular universe, how would we know that the process of finding it had not "cut out" the singularity in this artificial way?

The answer to this dilemma is to give up the traditional notion of the singularity as a place of infinite density and temperature. Instead, we say that a singularity occurs when the path through space and time of any light ray comes to a full stop and cannot be continued. What could be more "singular" than this Alice-in-Wonderland experience? At the end of its path, the light ray has reached the edge of space and time. It "disappears" from the universe. The elegant feature of this way of defining a singularity is that if the density does indeed become infinite somewhere, then the light ray's path will be brought to a halt, because space and time are destroyed. But if such a point has been excised from the universe, the light ray will also come to a halt when it reaches the perimeter of the hole that remains (see figure 3.2).

This picture of a singularity as the edge of a universe is

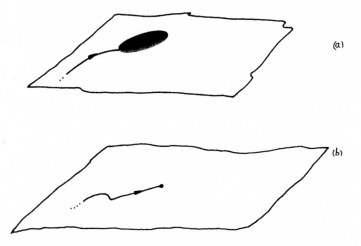

FIGURE 3.2
Two sheets representing universes of space and time in which the paths of light rays come to an end. In (a), a hole has been cut out of the universe and the light ray hits its boundary. In (b), the light ray hits a singularity, where space and time are destroyed.

an extremely useful one. It bypasses the problems that were raised about the shape and pressure of the universe, and the ambiguities of mapping it with coordinates. While such a singularity might be accompanied by extremes of density and temperature, as envisaged in our intuitive picture of an expanding, big-bang universe, it need not be.

There are other changes to our commonsense notions about the beginning of the universe. It need not occur everywhere at the same time. Different paths through time might be found to begin at different moments, if traced back to their singular beginnings. Perhaps the fact that some regions of the universe today are less dense than others is a reflection of the fact that they emerged from the singularity slightly earlier and have had longer to expand and rarify than the denser regions elsewhere.

During the mid-1960s, following the discovery of the microwave background radiation by Penzias and Wilson, the big-bang model began to be taken seriously, and cosmologists focused upon the issue of whether the universe had a singular beginning. Following the clarification of what this beginning should be identified with—inextendible paths backward through space and time—the challenge was to discover whether our universe contained a singularity of this sort, a beginning to time, in the past. Roger Penrose showed how such questions could be answered by novel geometrical arguments that astronomers had not previously made use of. Penrose's background in pure mathematics, and his remarkable geometrical intuition, equipped him to bring powerful new methods to bear upon the problem of discovering how light rays move and whether they can come from a past eternity or not. Later, he was joined in these efforts by Hawking and others, among them the physicists Robert Geroch and George Ellis.

Penrose showed that if the gravitational forces exerted

by material in the universe have always and everywhere been attractive, and if there is enough material in the universe, then the gravitational effect of that material makes it impossible for all light rays to be traced backward in time forever.

Some of them (maybe all of them) have to come to a full stop—a singularity—which we identify with our intuitive picture of the big bang (see figure 3.3). The beauty of these rigorous mathematical deductions is their avoidance of all the uncertainties about mapping coordinates and special symmetries. They do not require us to know lots of details about the universe's structure, or even to know the law of gravitation. It should be stressed that they are *theorems*, not theories. They lay out particular assumptions about the nature of the universe which, if true, guarantee a past singularity by logic alone. If those assumptions are found not to hold in our universe, we cannot conclude that there was no singularity—we cannot conclude anything about the beginning at all. The theorems no longer apply to our universe.

The two assumptions—that gravity is always and everywhere attractive and that there is enough matter in the universe—are fascinating because although expressed in mathematical language they can be checked by observation. Remarkably, the material requirement was met by the newly discovered microwave background radiation itself. This left only the requirement that gravity be always and everywhere attractive. In the 1960s, this was regarded as being an entirely reasonable assumption. There was no observational evidence against it and no well-founded theories of how matter would behave at high density which predicted that some forms of matter might antigravitate. In everyday circumstances, gravitational attraction is a consequence of matter having positive mass, and hence positive

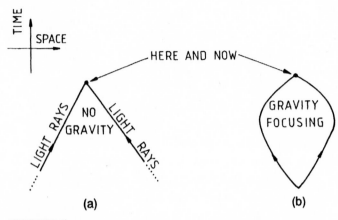

FIGURE 3.3

(a) The paths through space and time traced out by light rays moving at the (constant) speed of light in the absence of gravity. (b) Gravity bends the paths of the light rays from their straight-line paths. If there is sufficient matter in the universe, light rays will converge in the past to a singularity.

··········

density. But when one deals with material compressed to very high densities, or moving at speeds close to the speed of light, c, one must recall Einstein's $E = mc^2$ formula again. Any form of energy, E, has an equivalent mass, m, and so it feels the gravitational force of other material. As noted, pressure is a form of energy (it arises from the energy of motion of molecules in a gas, for instance) and so is subject to gravitational force as well. Since there are three dimensions of space in which the particles creating the pressure can move, the stipulation that the gravitational force be attractive is that a quantity, D, equal to the density, d, plus three times the pressure, P, divided by c^2 be positive: $D = d + 3P/c^2 > 0$. This is true of all the familiar forms of matter in the universe—radiation, atoms, molecules, stars, rocks, and so on. Because of this, the idea that the universe had been shown to possess a beginning in time became a much publicized one in the late 1960s and throughout most of the 1970s. Much of the work of mathematical cosmologists centered on understanding what would have gone on very close to that singularity and on discovering what the most complicated singularities would have done to material in their vicinity.

An interesting sidelight on this deduction of a beginning to time is that it undercut the ancient idea of a cyclic universe that periodically contracts into a big crunch and then emerges into a new expanding phase. If we trace our history back to a singularity, then there was no "before." There is no way of tracing the universe's history back to some earlier contracting state: that idea had to stay in the realm of science fiction.

If the universe did begin at a singularity from which matter appeared with infinite density and temperature, then we are confronted with a number of problems in our attempts to push cosmology any further. "What" determines the sort of universe that emerges? If space and time do not exist before that singular beginning, how do we account for the laws of gravitation, or of logic, or of mathematics? Did they exist "before" the singularity? If so—and we seem to grant as much when we apply mathematics and logic to the singularity itself—then we must admit to a rationality larger than the material universe. Moreover, to understand the present state of the universe we seem required to do the impossible—that is, to understand the singularity. But the singularity was a unique event: how can it be amenable to the scientific method?

At first, cosmologists set about examining the two possible strategies that we described earlier: finding principles that might dictate what a singularity was like, or trying to show that it didn't matter—that the universe would end up looking pretty much as it does today regardless of how it began.

We have highlighted some of the things cosmologists have discovered about the universe and some of the questions they would like to answer. If we want to explain something about the present state of the universe—why, for example, galaxies have the shapes and sizes they do—we need to work backward in time, reconstructing the past history of the universe using our knowledge of how matter behaves under conditions of very high density and temperature. We would like to check our deductions against pieces of evidence left in the universe from past events; unfortunately, things are not so simple. The universe covers its tracks very effectively, and there are few pristine remnants of the distant past. But more fundamentally, we do not know all the ways in which matter can behave at extreme temperatures and densities. Experiments on Earth, limited by economic realities as well as by constraints of size and available power, are unable to simulate fully the conditions that would have obtained in the universe during the first hundredth of a second of its expansion history.

This creates a fascinating state of affairs. The cosmologist looks to the elementary-particle physicist to provide an account of how matter and radiation behave at very high temperatures, so that the past history of the universe can be reconstructed ever closer to its apparent beginning. The particle physicist, on the other hand, cannot do this with the resources available on Earth. Terrestrial particle accelerators cannot reproduce the energies of the big bang, nor can their detectors catch the most ethereal elementary particles of matter. So particle physicists look to the early moments of the universe as a way of testing their theories. If their latest theory predicts that stars or galaxies cannot exist, then it can be excluded. However, one can see that a delicate balancing act is being performed, as partly tested (or even untested) physics is being used in order to draw up possible histories of the first second of the universe's history.

The reader would do well to regard the first second after the big bang as a cosmic watershed. It is believed that after this time the temperature in the universe was low enough so that terrestrial physics applies and can be tested experimentally. But our inability to re-create fully the physical processes and elementary particles that dictated the course of the universe in the first second makes our reconstruction of its history uncertain. One second is also the time at which conditions in the early universe determined the universal abundance of the element helium. Its abundance gives us a direct probe of the way in which the universe was expanding at that time.

This does not mean that we understand all the events that occurred after the universe was one second old. We understand the general physical principles and laws governing the behavior of the contents of the universe from that period onward, but there are sequences of events—particularly those associated with the formation of galaxies—which are of enormous complexity and which we have yet to reconstruct in detail. It is rather like our knowledge of weather systems. We know all the principles of physics that govern the weather, and we can account for any past sequence of climatic changes. But we cannot necessarily *predict* the weather, even tomorrow's weather, because of the complicated and sensitive interplay of countless factors that combine to determine the weather's current state. Because we cannot know that state perfectly, our ability to predict is limited.

During the late 1970s, the study of the most elementary particles of matter became linked to the sciences of astronomy and cosmology. Often, there would be astronomical consequences if some new variety of subatomic particle existed, even if its effects were too weak for it to show up in particle-collider experiments. In this way, one can use the astronomical evidence to rule out the existence of many new types of elementary particle.

A nice example of the symbiotic relationship between cosmology and the study of elementary-particle physics was provided by the interplay between the results of high-precision experiments at CERN (the European Center for Nuclear Research), in Geneva, and cosmological theories of nuclear reactions during the first few minutes of the universe's history. Each approach tells us how many varieties there are of an elementary particle called the neutrino. Neutrinos are ghostly particles that interact so weakly with all other forms of matter that they are very hard to detect; indeed, many are passing through your body at this very moment. Two varieties of neutrino—the electron neutrino and the muon neutrino—have long been known to physicists, and both have been detected directly in countless accelerator experiments. A third variety, the tau neutrino, reveals its existence only indirectly, through the decays of other particles; its production requires too much energy for it to have been detected directly as yet. Can we therefore be sure that the tau neutrino exists, and are there any other neutrino types we haven't seen yet?

First, let's see how our reconstruction of the history of the universe allows us to use astronomical observations to determine the number of neutrino varieties. Then we can compare the result with the recent CERN experiments, which measure this number directly.

Cosmologists have, since the 1970s, assumed that there are three and only three varieties of neutrino and used this input as part of the specification of the preferred model for the constituents of the universe in its youth. It is very important for them to know how many varieties of neutrino exist in nature, because this fixes the total density of radiation and matter during the very early universe, which in turn determines how fast the universe expands. They use all this information to investigate the detailed goings-on in the universe when it was between one and a thousand seconds old. During that niche in cosmic history, the expanding universe was hot enough for nuclear reactions to make the lightest elements, by fusing neutrons and protons together in different combinations. At earlier times, the temperature was so high that any element heavier than hydrogen, which consists of a single proton, would be broken up as soon as it formed (the hydrogen nuclei disappear as well, when the universe is less than a microsecond old). During the first ten seconds, the build-up of light elements is slow, because of breakups, but it climaxes in a frantic rush of nuclear activity after a hundred seconds, before rapidly being shut down by the falling temperature and density. After a thousand seconds, it's all over.

In order to predict the outcome of these nuclear reactions, one needs to know the relative numbers of protons and neutrons available. These numbers will determine the final abundances of the nuclei that are built from them: deuterium, an isotope of hydrogen, with one proton and one neutron; helium, comprising two isotopes, one with two protons and one neutron (helium-3) and the other with two protons and two neutrons (helium-4); and lithium, consisting of three protons and four neutrons.

When the universe is younger than one second old, there should exist equal numbers of protons and neutrons, because of the so-called weak interactions between them, which turn one into the other and keep their numbers in balance. But when the universe is one second old, the rate of expansion becomes too great for these weak interactions to maintain a perfect neutron-proton balance. It becomes slightly harder to turn a proton into a neutron than vice versa, because the neutron is a little heavier than the proton and its production therefore requires more energy to achieve. The weak interactions shut down, leaving a definite abundance of protons relative to neutrons: the ratio is seven to one. About one hundred seconds later, nuclear reactions start up which combine these neutrons and protons into nuclei of deuterium, helium, and lithium. Some 23 percent of all the material ends up as helium-4. Almost all the rest remains as hydrogen, with a few parts in a hundred thousand residing in the isotopes helium-3 and deuterium, and a few parts in ten billion as lithium (see figure 3.4).

Astronomical observations of helium, deuterium, and lithium around the universe confirm the existence of universal abundances of these magnitudes. There is a beautiful agreement between the simplest big-bang model and astronomical observations. It became clear that this agreement hinged on the assumption that there are only three varieties of neutrino in nature. If there were four, the expansion rate of the early universe would be increased, and more neutrons would remain relative to protons when the weak interactions shut down; there would thus be a corresponding increase in the final abundance of helium that emerged from the early universe. Very detailed studies were performed taking into account all the observations and their uncertainties. Cosmologists claimed that there could not exist another variety of neutrino similar to the three we know about (see figure 3.5).

The CERN experiments confirmed this prediction. They produced very large numbers of short-lived particles called Z bosons. Each is about ninety-two times as massive as a proton, and rapidly decays into lighter particles, including neutrinos. The more varieties of neutrino there are, the

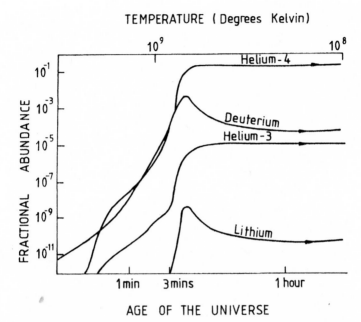

FIGURE 3.4

The detailed production of the lightest elements out of protons and neutrons during the first three minutes of the universe's history. The nuclear reactions occur rapidly when the temperature falls below a billion degrees Kelvin. Subsequently, the reactions are shut down, because of the rapidly falling temperature and density of matter in the expanding universe.

· · · · · · · · ·

more avenues the Z bosons have for decay and the faster they will disappear. The CERN experimenters monitored the decays of large numbers of Z bosons in order to determine how many varieties of neutrino they were decaying into. The answer was 2.98 ± 0.05, allowing for the uncertainties in the experiments. There appear to be only three neutrino types.

This is a nice example of how particle physics and cosmology complement each other and enrich our understand-

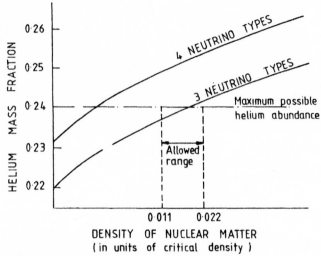

FIGURE 3.5

The abundance of helium-4 produced in the early universe for different values of the universal density of nuclear matter in units of the critical density needed to "close" the universe. The amount of helium produced is shown when there exist three or four types of neutrino. The observed fraction of the universe's mass in the form of helium-4 lies in the range 0.22 to 0.24. When the density of matter lies between 0.011 and 0.022 of the critical density, the abundances of helium-3, deuterium, and lithium-7 also agree with observations. This density range is also in accord with the observed density of matter found in stars and galaxies today. The existence of four neutrino types predicts far more helium than the maximum allowed (0.24). There is agreement between the observations and the predictions only when there are three neutrino types, in which case the predicted helium fraction lies between 0.235 and 0.240.

....

ing of the universe as a whole. The correct prediction of the abundances of the lightest nuclear elements is the greatest success of the big-bang picture of the universe. The predictions are sensitive to small changes in the structure of the universe when it was a second old. This enables us to draw conclusions about what the universe could have been like then. If, for example, it expanded at different rates in different directions, or contained strong magnetic fields throughout space, then the expansion rate is increased and the abundance of helium will be greatly in excess of what is seen. Astronomical observations of the lightest nuclear elements in space reach back deeper into the past than observations of the cosmic background radiation; they are therefore our most powerful probe of what the universe could have been like after it had been expanding for just a second.

There is one further and important feature of these studies of primordial nuclear reactions which exhibits a general feature of the big-bang model. Calculating the abundances of the elements formed in the early universe does not require information about what the universe was like at the beginning. The relative abundances of protons and neutrons are determined by the temperature of the universe when the weak interactions between them shut down. This is a remarkable feature of a big-bang universe; the hot state of equilibrium insures that the temperature determines the relative abundances of different particles of matter and radiation precisely. This fact was not fully appreciated until 1951. Before then, many cosmologists thought that the abundances of the elements in the very early stages of the universe depended upon the relative numbers of protons and neutrons present at the beginning. But this is not so. Before the universe was a second old, the numbers of protons and neutrons were equal. Some things are as they are regardless of what they were.

....

CHAPTER 4

INFLATION AND
THE PARTICLE PHYSICISTS

It has long been an axiom of mine that the little things are
infinitely the most important.
—*A Case of Identity*

In the mid-1970s, cosmology took a new direction. In 1973,
the particle physicists unearthed a successful theory of
how matter behaved under extreme conditions. Previously,
they had expected that its interactions would become
stronger and more complicated as energies and temperatures increased. Hence, their interest in studying the environment provided by the first second of the big bang was
less than enthusiastic; the soluble problems were more
pressing. But their successful new descriptions of high-
energy interaction between elementary particles showed
these interactions becoming weaker and simpler as temperatures and energies rose. This property was called
"asymptotic freedom," because if the energies were extrap-
olated to become infinitely great, then in this asymptotia
the particles would not interact at all.

Elementary-particle physicists had already begun to
search for ways of joining together the four fundamental
forces of nature—gravity, electromagnetism, and the strong
and weak nuclear forces—into a single unified theory. The

theory of how the weak force (manifested in a certain kind of radioactivity) and the electromagnetic force were entwined was first discovered in 1967 and would eventually be spectacularly confirmed by the discovery at CERN in 1983 of two new types of elementary particle that the "electroweak theory" predicted should exist. Now the search was on for ways to add the strong force (which binds the nucleus together) in order to produce a "grand unified theory" that would lack only gravity.

At first sight, these attempts at unification appear doomed, because we know that the fundamental forces of nature are of very different strengths and act upon different collections of particles. How could these disparate things be the same? The answer is that the strengths of the forces of nature vary with the temperature of the environment. So although they are very different from one another in the low-energy world in which we live, they will slowly change as we probe conditions of higher temperature. The prospective theories that were developed predicted that the strong force and the electroweak force should become of roughly equal strength at very high energies—about 10^{15}GeV, corresponding to temperatures of about 10^{28} degrees Kelvin—far in excess of those that could ever be created in any conceivable terrestrial particle collider but equal to the energies experienced in the early universe a fraction of a second (10^{-35} seconds) after its apparent beginning. So we might be able to test whether any grand unified theory made physical sense by exploring its cosmological consequences. Moreover, cosmologists might find that these new predictions about the behavior of elementary particles shed light upon unexplained properties of the universe.

As noted, the grand unified theories overcame the problem of unifying forces of different strengths by taking into account variations in their strengths with increasing temperature (see figure 4.1). The other problem they had to overcome was the fact that each force acted upon different classes of elementary particle. To fully unify the forces, these particles all had to be able to transmute into one another. This required the existence of intermediaries with very large masses—masses so large that the intermediaries would appear in profusion only when the universe was hot enough for particle collisions to create them. Two new varieties of heavy particle were predicted to arise quite inevitably in theories of this sort. The first, which we call the X particle, seemed to be a godsend: unlike any known elementary particle, it could change matter into antimatter. This feature enabled the grand unified theories to explain a curious lopsidedness in the universe.

Every variety of elementary particle in nature, except the photon, possesses an antiparticle, whose attributes have opposite values—analogous to the way that the North pole of a magnet is opposite to its South pole. Although laboratory experiments in particle physics produce particles and antiparticles entirely democratically, when we look out into space or collect cosmic rays we find evidence only for extraterrestrial matter—never for extraterrestrial antimatter. The universe seems to be dominated by matter—and if this is so today, cosmologists concluded, it must have been so at the beginning, since there seemed no way for antimatter to turn into matter. In other words, there must have been an initial asymmetry to explain the present imbalance; but explaining the present asymmetry by positing a special initial asymmetry does not seem to really explain anything at all. The only "natural" initial state, one might imagine, appears to be one in which there are equal amounts of matter and antimatter; however, such a state appeared unable to change into the lopsided one we observe today. This is

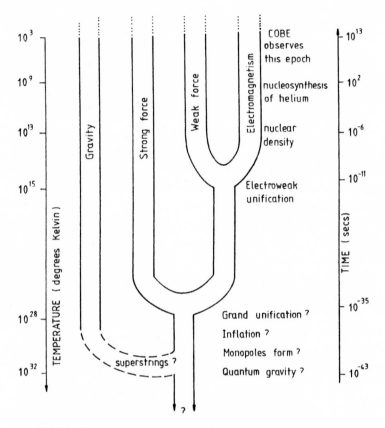

FIGURE 4.1

The expected temperature history of the first million years of the universe, as one goes backward to earlier and earlier times. As the temperature increases, the effective strengths of the fundamental forces of nature change, and unifications are expected to occur. These are shown by mergers of the forces.

where the X particles required by the grand unified theories came to the rescue. As a by-product of their mediation of the unification of the strong and electroweak forces, X particles allow matter to turn into antimatter. Yet X particles and their antiparticles do not decay at the same rate; consequently, an initial state that possessed a perfect balance between matter and antimatter (equal numbers of Xs and anti-Xs) could change into a lopsided one, by means of these decays, during the very early moments of the universe.

This possible solution to the observed matter-antimatter asymmetry provoked huge interest in the study of the very early universe among particle physicists during the period from 1977 to 1980. But there was also some bad news, to which people were largely turning a blind eye. Remember that the X particle was one of two types of particle inevitably produced all over the universe during its first moments. Whereas the X particles soon decayed into quarks and electrons, which lie within the atoms around us today, the second sort was unwanted and would not go away.

These unwanted particles, called "magnetic monopoles," are required by any grand unified theory if it is to produce a world like our own—one that contains the familiar forces of electricity and magnetism. Because of this link to electricity and magnetism, they could not be eliminated from the theory by tinkering with its structure. Instead, some way had to be found to remove them from the early universe once they had formed, because there is no observational evidence that they exist today. Worse still, if they did persist they would end up contributing a billion times more to the density of the universe than all the ordinary matter in stars and galaxies. That is not the universe we live in. For such a preponderance of matter in any form would have caused the universe to slow down and collapse to a big crunch billions of years ago. Neither galaxies nor stars nor people could exist. The problem was very grave. How could one get rid of these unwanted monopoles or suppress their production? The answer would open a new chapter in our thinking about the universe and completely change our approach to understanding how it might have originated. To understand the profundity of this change we need to embark upon an exploration of whether the universe we see today is all there is and why its present form is so mysterious.

When we talk about the universe, we must draw an important distinction. There is the universe—everything that is. It might be infinite in extent: it might be finite. We just don't know. Then there is something that we should call the *visible universe*—that finite part of the universe from which there has been time for light to reach us since the universe first began expanding (see figure 4.2). We can think of the visible universe as an imaginary ball about fifteen billion light-years in radius, with ourselves at the center. As time goes by, our visible universe increases in size.

Now suppose we retrace the history of the region that constitutes our visible universe today. It has been taking part in the universal expansion, so the material it contains (enough to make one hundred billion galaxies today) was contained within a much smaller region in the past. As the radius of this region increases with the expansion, the temperature of the radiation within it falls inversely with its size, in accord with the well-known and tested laws of thermodynamics. This means that we can use the radiation temperature as a gauge of the size of the currently visible part of the universe in the past. If it doubles in size, its temperature halves.

Let us now pick some very early time, and we shall make it the time at which the grand unification of three of the fundamental forces was predicted to have occurred.

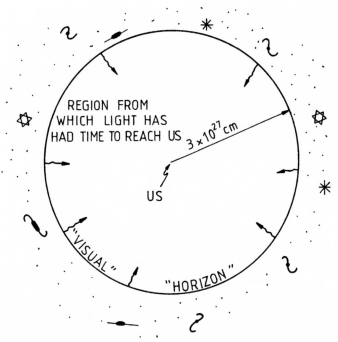

FIGURE 4.2

The visible universe is defined as that spherical region around us from which light has had time to arrive since the universe began expanding. The radius of this sphere is 3×10^{27} centimeters today.

··········

This is the epoch when the temperature of the universe was great enough to produce the X particles and monopoles. It corresponds to about 3×10^{28} degrees Kelvin, and we have to go back to just 10^{-35} seconds after the expansion began to find such an environment.

Today, after about 10^{17} seconds of expansion, the temperature of the radiation has fallen to 3°K. So the temperature has changed by a factor of 10^{28} since that early time, and the contents of the visible universe today were then contained within a sphere 10^{28} times smaller than the visible universe today. The radius of the visible universe today is given by its age multiplied by the speed of light. As noted in figure 4.2, this radius is roughly 3×10^{27} centimeters. Hence, at the epoch of grand unification everything within our visible universe was contained within a ball of three millimeters in radius! This sounds amazingly small, but the problem is that it is actually so *large*. For, at this time the distance that light can have traveled since the expansion began is the speed of light, 3×10^{10} centimeters per second, times the age 10^{-35} seconds, which gives 3×10^{-25} centimeters (see figure 4.3). This is the greatest distance over which any signal can have traveled since the expansion began. It is called the "horizon distance." If any irregularities in the initial state of the universe are being ironed out by friction or other smoothing processes, the horizon dictates the maximum extent of the smoothing at any time, because these processes cannot act faster than the speed of light. The problem is that the region that is going to expand to become our visible universe today was, at that early time, fantastically *larger* than the size of the horizon distance. This creates a puzzle and a problem.

The *puzzle* is how to explain the remarkable regularity of our universe from place to place and from one direction to another on the sky if it is made up of a huge number of separate regions that were once completely independent of each other—in the sense that there had not been sufficient time since the beginning of the universe for light to travel from one to the other. How did they come to have the same temperature and expansion rate to better than a part in a thousand (as the isotropy of the microwave background radiation demonstrates) if there was insufficient time for any transfer of heat or energy to coordinate them? We seem to be left with the conclusion that the initial state was such that conditions were simply "created" the same everywhere.

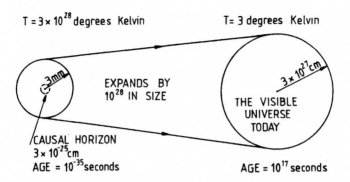

FIGURE 4.3

If we retrace the expansion history of our visible universe, it will be compressed into a region 3 millimeters in radius when it was 10^{-35} seconds old. But at this early time light has traveled only about 10^{-25} centimeters. This defines the "causal horizon" at this time.

· · · · · · · · · ·

The *problem* is the ubiquity of the magnetic monopoles. These particles arose in the early universe where there were mismatches in the orientations of unusual energy fields. Wherever there was a mismatch of the directions in which these fields point, a knot of energy—a monopole—formed. The horizon diameter at this time, 10^{-25} centimeters, tells us the range over which the directions of those energy fields can be aligned and mismatches avoided. But at that very early time the region that would expand to become our visible universe was 10^{24} times bigger than the horizon size, and so should have contained a huge number of mismatches, resulting in an unacceptably large number of monopoles today in our visible part of the universe. This is called the "monopole problem."

It is useful to step back from these details and take stock of what has happened. Physicists have devised detailed theories of how matter behaves at very high temperatures.

Such theories ought therefore to find application to the first moments of the universe's history. When they are used to reconstruct those first moments, they yield exciting new insights, explaining, for instance, how the universe came to favor matter over antimatter. But they also predict the existence of a huge cosmic abundance of new particles of matter called magnetic monopoles—an abundance that does not exist. The reason that so many monopoles are expected is that the entire visible universe today has expanded from a region that, at the time the monopoles were produced, was far larger than the distance that light can have traveled since the expansion began, and so would have contained very many energetic mismatches of the sort that create monopoles. Physicists were so impressed by the successes of these grand unified theories that, rather than abandon them in the face of the monopole problem, they put that problem to one side and continued exploring the other properties of the theories, hoping, like Mr. Micawber, that something would turn up. It did.

In 1979, Alan Guth, a young American particle physicist working at the Stanford Linear Accelerator Center, hit upon a way to solve this problem and make the idea of grand unification compatible with what we know about the universe. Since that time, his concept of the "inflationary universe" has become a focal point for studies of the very early universe, and inflation theory has blossomed into a discipline in its own right, examining all the ways in which his basic idea can be realized.

We can see that the monopole problem is a consequence of the small size of the horizon in the very early universe. By the present day, the horizon size at the epoch of grand unification would have expanded into a region no more than one hundred kilometers across. If only the universe had expanded faster in its early stages, we might be able to expand the horizon-size region into one the size of the visible universe today. This is what Alan Guth's inflationary-universe hypothesis proposes. It requires the universe to undergo a brief period of *accelerated* expansion during its very early stages. The period required is very short—acceleration from just 10^{-35} to 10^{-33} seconds will do the trick (see figure 4.4).

If this acceleration occurs, the whole of our visible universe can expand from the region small enough to have been traversed by light signals in the time since the expansion began. Its smoothness and isotropy thus become understandable. But most important, there will not be large numbers of monopoles, because our visible universe will have expanded from a region so small that it could encompass, at most, *one* of the mismatches that produces them. The monopole problem would be solved. Likewise, the observed uniformity of the universe is explained not by some new mechanism for eradicating it, or by some principle demanding that the initial state be meticulously orderly, but by the fact that we see only the expanded image of a

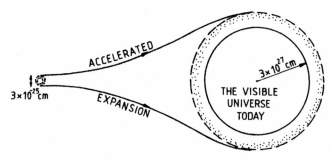

FIGURE 4.4

Inflation accelerates the early expansion of the universe, enabling the region of radius 10^{-25} centimeters to expand to the scale of our visible universe today. Compare this with the slower early expansion history of figure 4.3.

region small enough to be kept uniform initially by smoothing processes that carry excess energy from hotter regions to cooler ones. The nonuniformities might still be out there, beyond our present horizon. They have not been dissipated—merely swept where they cannot be seen.

The fleeting period of cosmic history during which the expansion of the universe accelerates sounds like a minor gloss on the universe's history, but it has momentous and far-reaching implications. Earlier, we discussed the singularity theorems of Penrose, Hawking, Geroch, and Ellis; you recall that they hinged upon the assumption that always and everywhere matter experiences gravitational attraction to other matter. On page 44, we described how the attractive force of gravity requires the quantity D, a sum of the density and pressures in the universe, to be positive. When this is the case, all expanding universes decelerate, and this was the expectation for all big-bang models before inflation was proposed. Regardless of how fast they begin expanding, and whether they will expand forever or collapse to a big crunch, the effect of gravity is to slow down the expansion, because of the attraction that all matter exerts upon other matter. So if one wants the early universe to experience a brief period of accelerated expansion, it is necessary for the effects of gravity to become temporarily repulsive and hence for the quantity D to become temporarily negative. This is the heart of the inflationary-universe hypothesis: it is an explanation of the uniformity of the universe and a resolution of the monopole problem, based upon the requirement that antigravitating states of matter exist that can create this brief period of acceleration soon after the big bang. If no such material exists in nature, the theory fails. It it does exist, then in the next chapter we shall see that some fossil evidence should remain in the universe as a witness to that past era of inflation.

During the 1960s, it was believed to be entirely reasonable that all forms of matter would exhibit gravitational attraction rather than repulsion. During the 1980s, however, cosmologists came to believe that matter at high density experiences conditions in which gravitational repulsion will occur. What has produced this volte-face has, again, been the opening up of new possibilities by particle physicists, whose theories predict new forms of matter that can create very large negative pressures. These negative pressures can be sufficient to outweigh the positive density and produce gravitational repulsion (that is, a negative value of the quantity D, on page 44). If these forms of matter exist in reality and not just on paper, they can increase in strength very slowly as the universe expands and eventually exert their antigravitational effects upon the expansion. The expansion will thus begin to accelerate. The universe will "inflate" until the matter-fields that are responsible decay into more mundane forms of matter and radiation—forms that display only gravitational attraction. The expansion will then revert to the decelerating state it had before inflation began, and which it has today. This is the essence of the inflationary-universe scenario for the evolution of the very early universe (see figure 4.5).

The attraction of this picture of cosmic history for cosmologists is manyfold. We have seen how it resolves the monopole problem and allows us to understand the uniformity of the universe over its largest observed scales. But it makes two further predictions about the present state of the visible universe which will allow us to rule it out if it is wrong.

In order to solve the monopole problem, the period of accelerated expansion must last at least seventy times as long as the age of the universe when acceleration begins. This is what it would take to produce our visible universe

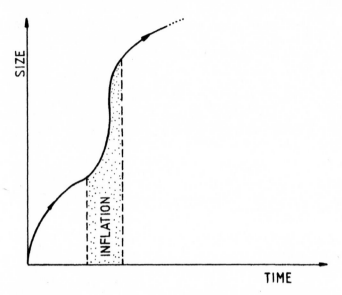

FIGURE 4.5

The variation of the radius of an inflationary universe with time. The period of inflation is greatly enlarged in the picture. In practice, it need only last from 10^{-35} seconds to 10^{-33} seconds after the expansion begins. The present age is about fifteen billion years. The figure shows how the expansion of the universe begins in a decelerating fashion, then accelerates for a period when inflation occurs, before returning to a state of decelerating expansion after inflation ends.

• • • • • • • • • •

from a region one millimeter across early on. An important consequence of this accelerated expansion is that it makes the universe expand faster and for a longer time than it otherwise would have. Without inflation, it would have expanded naturally only for a fraction of a second before contracting; with inflation, the expansion can easily last for more than trillions of years. The acceleration drives our expanding universe very close to the critical divide that separates universes that will expand forever from those doomed ultimately to contract back to a big crunch. Inflation therefore provides a natural explanation for the observed mysterious proximity of the visible universe to the critical divide (see figure 4.6).

If the period of accelerated expansion lasts long enough to explain why we do not see any magnetic monopoles, we should find the present expansion to lie within one part in a million of the critical divide—that is, we should find the average density of the visible universe to lie within one part in a million of the critical value, which is 2×10^{-29} grams of matter per cubic centimeter of space, on the average.

This is interesting for two reasons. First, if the density

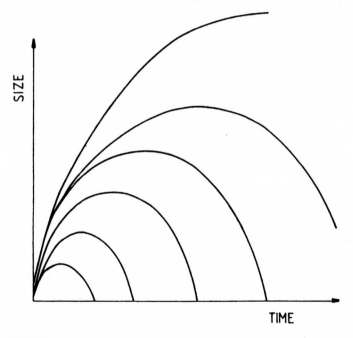

FIGURE 4.6

A collection of different closed universes with different total lifetimes. The universes that expand longest get driven closest to the critical divide.

is that close to the critical divide, we will never be able to determine whether our universe is open or closed: our observations cannot measure the density of the visible part of the universe to an accuracy of one part in a million. But of more immediate interest is the second implication—because the observed density of *luminous* material is at least ten times as small as the critical level. If the inflationary theory is correct, most of the material in the visible part of the universe must exist in some nonluminous form, rather than in shining stars and galaxies. This is a welcome conclusion, because astronomers have long been puzzled by the fact that observations of the motions of stars and galaxies indicate that these objects are moving faster than they can be moved by the gravitational forces that would be exerted by the luminous material nearby. Apparently, a great deal of dark, unseen material exists, whose gravitational pull is responsible for the motions of the stars and galaxies that we see.

Our first reaction to this bookkeeping is to suppose that there must be a lot more dark material out there in between the stars and galaxies (perhaps in the forms of very faint stars, or rocks, gas, dust, and other bits of debris)—material that did not get incorporated into the processes that led to stellar formation. This means that a map of the light distribution in the universe is not a good guide to the distribution of matter—a not unfamiliar situation. If we were to look at the Earth from space and draw a map of the distribution of the nighttime lighting, we would not find that it mapped the population density very faithfully. Rather, it would mirror the distribution of wealth. Great Western cities would shine out brightly, but the vast population centers of the Third World would be dim.

Unfortunately, things do not appear to be so straightforward in the universe. While we might think it reasonable for the universe to contain large amounts of ordinary atomic and molecular material scattered around in nonluminous forms, nature does not appear to share our opinion. You remember that one of the cornerstones of our theory of the expanding universe was our ability to predict in detail the result of the sequence of nuclear reactions that should have occurred by the time the universe was just a few minutes old. Those computations produce a remarkable agreement with the observed abundances of hydrogen, lithium, deuterium, and the two isotopes of helium. They tell us that the density of matter which takes part in those nuclear reactions must contribute no more than a tenth of the critical density. If the actual density were greater than this, the nuclear reactions would have incorporated so many of the neutrons into helium-4 that far fewer deuterium and helium-3 nuclei than we observe today would be left behind as by-products. The abundances of helium-3 and deuterium act as a sensitive measure of the cosmic density of nuclear matter. They tell us that if the universe has a near-critical density of dark material hidden away within it, this material cannot be in any form that takes part in nuclear reactions.

This means that it must be in the form of neutrinolike particles. Neutrinos, it will be recalled, carry no electric charge and so are unaffected by the electromagnetic force. Nor do they feel the influence of the strong nuclear force; they feel only the effects of gravity and the weak force.

We know of three different varieties of neutrino, and none has ever been found to possess a mass different from zero. But such evidence is not very strong; because neutrinos interact so weakly, the experiments to detect their masses are extremely difficult and rather insensitive to the tiny mass that a neutrino might possess. But particle physicists have more to offer us. Their attempts to unify all the forces of nature predict the existence of massive, weakly interacting particles called WIMPs (an acronym for *W*eakly *I*nteracting *M*assive *P*articles), which we have not yet detected in terrestrial experiments. One of the aims of a new particle collider planned in Geneva is to discover some of these massive particles.

If the three known varieties of neutrino possess masses that add up to no more than about ninety electron volts (an atom of hydrogen has a mass of about a billion electron volts), all those neutrinos spread throughout the universe will contribute a density that exceeds the critical value, and the universe will be "closed"—that is, destined to collapse in the future. Similarly, if WIMPs exist and have masses just twice that of the hydrogen atom, then the big-bang theory predicts that their cumulative density will equal that required to close the universe.

If the universe is primarily composed of a sea of these weakly interacting particles, we might ask why we cannot detect them directly and settle the matter once and for all. Unfortunately, we have no hope of ever directly detecting a universal sea of the known neutrinos; because of their tiny mass, they interact too feebly with our detectors. All we can do is attempt to measure the mass of the neutrinos in the laboratory—where they can be produced with energies that allow their effects on other particles to be observed—and test our expectations of what they will do to the clustering of luminous matter by comparing computer simulations of the clustering process with observations. However, if it is the WIMPs that constitute the dark matter, things are much more exciting. These particles are a billion times more massive than the known neutrinos could be, and they should hit our detectors with much more energy. In fact, it is just within our capabilities to

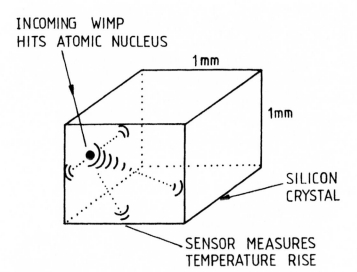

FIGURE 4.7

A physical process that can be used to detect WIMPs. A small crystal (1 millimeter along each side) is cooled to a few hundredths of a degree Kelvin above absolute zero. An incoming WIMP hits the nucleus of an atom in the crystal. The nucleus recoils but is quickly slowed down, giving out the energy from the recoil as blast waves, which heat up the crystal by a small but measurable amount.

• • • • • • • • • •

detect a sea of these particles in the universe around us, if they are prevalent enough to constitute the dark matter.

At present, several experimental groups in Britain and the United States have constructed underground detectors in an attempt to discover a cosmic sea of WIMPs. When one of these particles hits an atomic nucleus in a crystal, it will cause the nucleus to recoil and register a signal by slightly heating the crystal with the energy deposited. If one monitors just a kilogram of material for these events, one should find between about one and ten of them every day. If you can shield out all the other signals—from cosmic rays, radioactive decays, and other terrestrial events—that would otherwise swamp the detector, it should be possible to determine whether WIMPs are all around us. This shielding is achieved by placing the detector deep underground, inside a refrigerator that cools it to within a degree of absolute zero, and surrounding the refrigerator with absorbent materials (see figure 4.7).

Over the next few years, we hope to see the first results from experiments like these. They promise to unveil remarkable things about the universe in unexpected ways. Whether the universe is open or closed may hinge upon the properties of the smallest particles of matter and may be decided at the bottom of mineshafts on Earth rather than with telescopes focused upon the heavens. The great clusters of galaxies may be but a drop in the ocean of matter in the universe. The bulk of that matter, perhaps enough to curve space up into closure, may be in a form quite unlike any we have yet detected in our particle accelerators. This would be the final Copernican twist to our status in the material universe. Not only are we not at the center of the universe; we are not even made out of the predominant form of matter in the universe.

....

CHAPTER 5

INFLATION AND THE COBE SEARCH

It is quite a three-pipe problem, and I beg that you
won't speak to me for fifty minutes.
—*The Red-headed League*

In the spring of 1992, the news media of the world became
excited by the announcement that NASA's Cosmic Background Explorer (COBE) satellite had observed tiny variations in the temperature of the microwave background
radiation around the sky. By observing the radiation from
above the Earth's atmosphere, the COBE satellite avoided
spurious variations created by atmospheric changes and
achieved greater accuracy than any similar Earth-based
experiment. What it did was continually to switch its
detector back and forth across the sky, over angles greater
than about ten degrees (for comparison, the face of the full
moon corresponds to about half a degree on the sky), and
determine the difference in the temperature of the
microwave background radiation photons coming toward
us from those directions. What do the tiny temperature
variations mean, and why did everyone get so excited
about them (some extravagant commentators going so far as
to claim the COBE data as the most important scientific
discovery of all time!)?

We can understand the existence of structures like stars and planets by using our knowledge of basic physical principles. But when it comes to galaxies, our understanding is much more uncertain. We do not know whether a similar strategy of identifying balances between different forces of nature will suffice to explain why galaxies and clusters of galaxies have their observed masses and shapes and sizes. Almost certainly it will not. Galaxies and galaxy clusters are islands of material where the density of matter is enormously greater than the average density in the outside universe. The average density of the Milky Way galaxy, for example, is about a million times greater than that of the universe. That such irregularities exist is not mysterious. If we take a perfectly smooth distribution of matter and introduce a tiny nonuniformity, it will snowball, growing more and more pronounced. For there will be a greater gravitational pull toward any place where there is a slight overabundance of matter, and yet more matter will be drawn into it at the expense of the sparser regions beyond, and so the buildup proceeds.

This process is called "gravitational instability," and it was first recognized by Isaac Newton three hundred years ago. Gravitational instability operates whether or not the universe is expanding—although aggregates of matter take longer to build up in an expanding universe, because the expansion tends to pull the aggregating material apart. But as the universe ages, the aggregates should become so dense compared with the rest of the universe that they cease to expand with the universal expansion (see figure 5.1). Instead, they become stable islands of matter, held in a balance between the inpull of gravity created by their constituents and the outward pressure exerted by the motions of their constituents. However, one can see that if we want to explain the origin of galaxies and galaxy clusters by the process of gravitational instability, we need to know something about conditions at or soon after the beginning of the expansion of the universe. The density level that those irregularities attain by a given time depends upon the density level they began with.

Our astronomical observations of the most distant galaxies and their suspected precursors reveal that over-densities like those we see today existed when the universe's expansion had reached only a fifth of its present extent. But what we really need is a look at how large the overdensities were when the expansion was only about a million years old and had reached only one-thousandth of its present extent. Conditions at this time, long before the overdensities looked anything like galaxies and galaxy clusters, are what the COBE satellite observed. The infor-

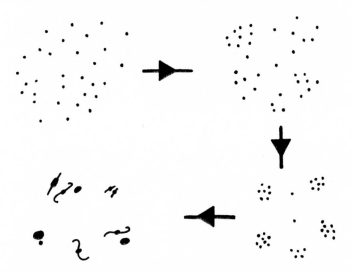

FIGURE 5.1
The process of gravitational instability gradually transforms a slightly nonuniform distribution of matter into one that is increasingly clumped.

mation it gathered was fossilized in the cosmic pattern of microwaves, untouched by subsequent events. The COBE data are now starting to be supplemented by the results of sensitive ground-based experiments.

As we have seen, radiation from the hot early stages of the universe cools as the universe expands. When the expansion has continued for about a million years, the radiation has cooled to such an extent that whole atoms and molecules begin to assemble out of the nuclei and electrons. At earlier times, they would have instantly disintegrated in encounters with the high-energy photons of radiation round about. At this time, the photons start to fly freely through space and time—carrying with them information about conditions where they originated—to become the microwave background radiation we observe today. In places where the density is a little higher than average, the temperature of the radiation will fall off a little more slowly than it will in sparser regions. What this means is that variations in the temperature of the microwave background radiation today provide us with a snapshot of the matter distribution in the universe when it was only a million years old—long before fully-fledged galaxies formed.

Cosmologists spent many years unsuccessfully searching for these variations with ground-based detectors; the COBE satellite finally found them. They were very small—only one part in one hundred thousand. This number tells by how much the nonuniformities need to amplify, by means of gravitational instability, in order to become strong enough to create the the first galaxies and clusters when the universe is billions of years old. It helps us begin to pin down the detailed events that led to the formation of galaxies in the intervening period. The discovery of these fluctuations in the background radiation was certainly exciting, but cosmologists were not really surprised. The surprise would have been the absence of such fluctuations, because then we would have had to assume that the galaxies had formed somehow without the benefit of initial irregularities—hence, not by the simple process of gravitational instability.

These fluctuation levels also give us a way of testing aspects of the inflationary-universe hypothesis. To see how this is possible we need to consider the phenomenon of inflation a little more closely.

Before the idea of inflation was introduced, the origin of galaxies and galaxy clusters was a largely intractable problem. There was no principle that could tell us how fluctuations in the density of matter and radiation arose in the first place, when they arose, or how big they were at the time of the decoupling of radiation from matter. All one could do was trace the present-day patterns of galaxies backward in time, assuming that gravitational instability had occurred, to determine the tiny level of irregularity required at any given earlier time. Unfortunately, the level of random fluctuations that would be expected to exist at any time in the universe was far too low to generate the structures we see today.

People soon realized that the inflationary idea offered a new solution to the puzzle. If a tiny region experiences a period of accelerated expansion, the random fluctuations, too, are inflated, and become the seeds of irregularities out to the scale of our visible universe today and beyond. The level of the fluctuations is determined by the antigravitating forms of matter (forms with negative D) that are responsible for the accelerated expansion. If one has a particular candidate for that material, one can predict the level of the fluctuations arising at the time of inflation. This is potentially a big step forward in the quest to understand the origin of galaxies and clusters. While we do not need to know how the universe began, we do need to know the particular variety of antigravitating matter that initiated inflation, because the level of the irregularity produced depends quite sensitively upon its identity and upon the strength of its interactions both with itself and with other, ordinary forms of matter. If inflation happened, the strength of the COBE signal tells us how strong those interactions were. Fortunately, there is more information in the COBE signal—information that does not depend so critically upon the identity of the antigravitating matter that drove the inflation.

When we map the distribution of galaxies and galaxy clusters in the universe, we find that the level of clustering depends upon the scale we are surveying. As we observe larger and larger chunks of the universe, we find the clustering steadily thins out, so when we talk about the level of irregularity in the universe we must specify the scale we are interested in. This variation with scale is called the "spectral slope" of the irregularity. It can be determined observationally—either by looking at the pattern of galaxy clustering or by observing the temperature variations in the microwave background radiation over a range of angles on the sky.

One of the appealing things about the inflationary theory is that it predicts that a particular spectral slope will most probably arise. The relative temperature variation—the temperature difference measured between two directions on the sky, divided by the average sky temperature—should not change as the angle between those two directions is increased. We call such a spectral slope "flat."

The COBE observations were of fundamental importance because they finally discovered evidence of the embryonic fluctuations that grew into galaxies and clusters. But for cosmologists the most interesting prospect is to see whether or not the spectral slope of the fluctuations accords with the predictions of the simplest inflationary theory. The COBE satellite gathered data in separate observing periods over several years, and the raw data require complicated processing to remove the known effects of the local environment—the satellite's electronics, the proximity of the moon and the Earth, and so forth— which create a statistical uncertainty in the result. The first round of observations were published in 1992, and tell us with 70 percent certainty that the spectral slope lies between -0.4 and +0.6. (The flat spectral slope would be zero.) When further COBE observations were processed early in 1994 and the original data were reanalyzed using more computer programs, all the data were found consistent with a spectral slope between -0.2 and +0.3, with 70 percent certainty. Further data analysis should narrow the interval in which the slope can lie. If those data home in on the zero value, it will provide a remarkable confirmation of the predictions of the simplest inflationary-universe models.

The COBE satellite could check the form of the spectral slope only by measuring the background-radiation temperature over angles of ten degrees or more. To scan smaller angles, a much larger experimental set-up than can be carried into space is needed. At present, there are a number of high-precision observations being conducted at sites on the Earth's surface: Owens Valley, in California; Tenerife, in the Canary Islands; and the South Pole. (The reason that the larger angular separations on the sky are not also probed from the ground is that the Earth's atmosphere varies too greatly over those separations, and the data are correspondingly affected.) In January 1994, the Tenerife team published evidence of temperature fluctuations over angles larger than four degrees. They report data consistent with the COBE observations, showing the spectral slope to be greater than -0.1.

To summarize our discussion: We find that the short period of accelerated expansion which we call "inflation" will necessarily produce tiny variations in the density of the universe from place to place—variations exhibiting a particular spectral slope. This spectral slope imprints itself upon the microwave background radiation, and we are able to check whether the spectral slope observed by the COBE satellite accords with the inflationary predictions. So far, it does. We therefore have a direct observational probe of physical processes that may have occurred when the universe was only 10^{-35} seconds old. We should consider our good fortune in this respect. There is no reason that the universe should be designed for our convenience. We wonder whether we will ever find all the laws of nature or have the wit to unravel the deepest mathematical structures at the heart of those laws by human thought alone. But suppose we did: it would then be remarkable good luck to find the means to test those ideas by experiment. Why should there be any relics from the first instants of the universe to allow us to check our ideas about what went on then? Vital observational facts about the deep structure and distant past of the universe are few and far between, but the wonder is not that there are so few such relics but that there are any at all.

We have glimpsed something of the idea of the inflationary universe and its observational consequences. As yet, this picture of how the universe expanded in its early moments remains under the heading of promising theory. Future processing of data from the COBE satellite and complementary ground-based experiments may indicate whether the predictions of the theory agree or conflict with the variations in the cosmic background radiation. But suppose, like optimistic theorists, we assume that inflation is the right approach to cosmology and pursue this approach until it is ruled out by observations. What implications does inflation have for our picture of the beginning of the universe?

First, we should recall that the condition for inflation to occur—the presence of forms of matter with negative D—is precisely the *opposite* of that assumed in the singularity theorems of Penrose, Hawking, Geroch, and Ellis. In a universe that inflates, these singularity theorems just don't apply, and we can draw no conclusions at all about a beginning to the universe. There might have been a singular beginning, but then again there might not have been. Yet despite this note of uncertainty, inflation can enlarge our conception of what the universe might be like in quite extraordinary ways.

When we discussed the onset of inflation, we described the process as if it had occurred in identical fashion everywhere in the universe. In reality, it would have occurred slightly differently from one place to the next. Suppose that the universe, in its preinflationary stage, was divided into regions, each of them small enough for light to have traversed them by the time inflation began. In each of these regions, the temperature and density will differ slightly (because of random fluctuations) or even dramatically (because of different starting states), with the result that the period of inflation will be of differing duration. One or another microscopic domain may inflate enormously, so that it will eventually become at least fifteen billion light-years in size, while in others virtually no inflation will occur (see figure 5.2).

We can imagine a chaotically random initial state to

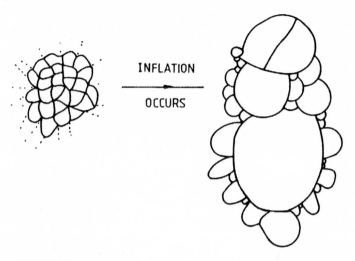

INFLATION

OCCURS

FIGURE 5.2
Chaotic inflation. Tiny regions of the very early universe experience different amounts of inflation. Only those regions that inflate enough to produce universes at least nine billion light-years in size will produce stable stars, carbon, and living observers.

· · · · · · · · ·

the universe—a universe that might even be infinite in extent. In some of the regions of space, conditions will permit inflation to occur by the amount required to produce a visible universe of the size we see today. In others, it will not. If we could see beyond the horizon of our visible part of the universe, we would eventually encounter some of those other inflated domains. They could have densities and temperatures very different from our own. When some inflationary-universe models are explored in this light, it emerges that there can exist even more radical differences: for example, the number of dimensions of space might vary from one part of the universe to another.

This model, known as the chaotic inflationary universe, was first suggested by the Soviet physicist Andrei Linde, in 1983. It introduces a new consideration into the study of the universe. We have already explained that the large size and great age of our visible universe is not coincidental; it is a necessary condition for the existence of biochemical complexity of the sort that we call life. Of all the microscopic domains that undergo differing amounts of inflation, only those that inflate by enough to grow to billions of light-years in size will produce stars—and thus the heavy elements necessary for biological complexity. We learn an important lesson from this insight. Even if it is highly improbable that any domains will experience such a large amount of inflation, we cannot exclude the scenario, because we can inhabit only such an improbably large domain. Moreover, if the universe itself is infinite in extent, domains of all sorts must exist, including some that inflate enough to produce a region like our visible universe.

Linde noticed that this chaotic picture of inflation had a further unexpected feature. Some inflating domains create internal random fluctuations that enable subregions of them to inflate, and these subregions, in turn, will produce subregions that can inflate further, and so on ad infinitum. Once inflation begins, it seems to be able to perpetuate itself. Beyond our horizon there must be regions still undergoing inflation. It appears that this perpetual inflationary process might not have had a beginning, but this remains an unresolved question (see figure 5.3).

These twin scenarios of chaotic and eternal inflation illustrate the ways in which the inflationary-universe idea enlarges our conception of space and time. They suggest that the universe is vastly more complicated than the small part of it we call "the visible universe." Before the introduction of inflation theory, such possibilities were dis-

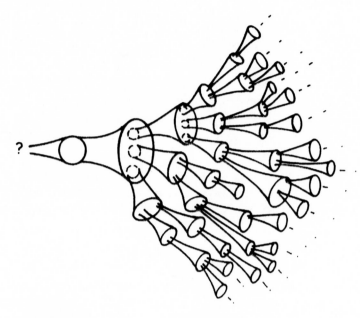

FIGURE 5.3
Eternal inflation. Each region that inflates creates conditions for subregions of it to inflate in turn, and so on ad infinitum.

· · · · · · · · · ·

cussed only as metaphysical speculations. The inflationary model, based as it is upon specific particle-physics models, turns these metaphysical constructs into possible consequences of entirely reasonable conditions in the early universe. Before inflation was proposed, we thought it most plausible to regard the visible universe as similar, on the average, to the rest of it. Now we do not.

Yet though inflationary cosmology offers fascinating possibilities, they are clouded by uncertainty. Inflation enables us to understand why the visible universe displays many of the properties it does, regardless of how the universe itself started out. This is a powerful feature: we can predict the present without having to know everything about the past. But there is a downside—the same downside noted at the end of chapter 1. If the present does not depend critically upon the details of how the universe began, we cannot deduce those details by observing the universe today. Inflation wipes the slate clean.

But what if inflation never happened? Alternatively, what if we focus upon the preinflationary history of only one of those inflating domains we have just described. What might we find if we follow it backward in time? Of course, we may still find a singularity of infinite density and temperature. But there are at least four quite different possibilities, all of which are consistent with everything we know about the universe (see figure 5.4).

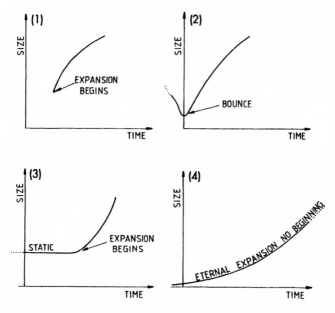

FIGURE 5.4
Some hypothetical beginnings to the expansion of the universe.

1) Instead of beginning as a state of infinite density, the universe of space, time, and matter comes into being with a finite density and continues in a state of expansion.

2) The universe "bounces" into a state of expansion from a previous state of maximum but finite contraction.

3) The universe suddenly begins its expansion from a static state, in which it has resided for a past eternity.

4) The universe gets ever smaller in the past without reaching a state of zero size. It has no beginning.

Why is our knowledge so uncertain? Why is it so difficult to extrapolate our theories back that last fraction of a second to decide whether or not they lead to a definite beginning. Earlier, we highlighted some crucial stages in the history of the universal expansion. After one second of expansion, conditions have cooled sufficiently to be described by terrestrial physics, and we have direct evidence left over from that time to check our reconstruction. Going back to just 10^{-11} seconds after the expansion began, we encounter conditions like those within the biggest particle colliders on Earth today. Before that time, we are beyond the range of conditions that we can partly simulate on Earth. Moreover, our knowledge of the laws of nature involved at these energy levels is also uncertain. For we are still building up a correct and complete theory of the elementary particles of matter, the forces that govern them, and what their effects upon the course of the universe's expansion will be. All this study is carried out on the assumption that Einstein's theory of gravitation correctly describes the expansion of the universe as a whole. True, it has passed all the observational tests ever set for it with amazing success. But it will not hold good all the way back to the beginning of the expansion. Just as Newton's description of gravity broke down when confronted with motions close to the speed of light and with very strong gravity fields, so we expect to encounter a regime in which Einstein's beautiful theory ultimately fails. This regime is the one that will be encountered if we try to probe the first 10^{-43} seconds of the expansion. At this "Planck time," as it is known, the entire universe is dominated by quantum uncertainty, and can be fully described only when we know how to unite gravity with the other three forces of nature in an all-encompassing "Theory of Everything." If we are to decide whether or not the universe had a beginning in any sense, we have to understand how gravity behaves in this period. That behavior is a manifestation of the peculiarities of the quantum aspects of matter.

The weirdness of the Planck time can be appreciated by examining the quantum picture of the subatomic world—a picture that has been built up in enormous detail over the last seventy years. It is the most accurate part of physics, and the technological marvels that surround us—from computers to CAT scans—are built upon quantum mechanics. When we try to observe things that are very small, the act of observation itself will significantly disturb the state we are seeking to measure. As a result, there is a fundamental limit to the accuracy with which the location and motion of something can be measured simultaneously. In the subatomic world, we cannot predict definite outcomes of measurements or other interactions—only the *probabilities* of observing particular outcomes. This state of affairs is usually described by noting that matter and light, which we tend to think of as tiny particles, exhibit the properties of waves under some circumstances. One can liken these "particle waves" to waves of emotion rather than water waves—that is, they are waves of information. If a wave of emotion sweeps through your neighborhood, it means that emotional behavior is more likely to be found there. Likewise, if an electron wave reaches your detector, it means that you are more likely to detect the electron there. Quantum mechanics tells us what the wave behavior of each particle of matter is, and hence the likelihood that one or another property will be detected.

Every particle of matter has a characteristic wavelength associated with its wavelike quantum aspect. This wavelength is inversely proportional to the mass of the object. When something is much larger than its quantum wavelength, one can for all practical purposes ignore the uncertainties introduced by its quantum nature. For large objects, like you and me, the quantum wavelength is very, very small, and we can safely ignore the wavelike uncertainty aspects in a car's position when we set out to cross the road.*

Suppose we apply these considerations to the visible universe. Today, it is vastly larger than its quantum wavelength, and we can neglect the tiny effects of quantum uncertainty when describing its structure. But as we go backward in time, the size of the visible universe at each past time is smaller, because the size of the visible universe when it has age T is the speed of light multiplied by T. The Planck time of 10^{-43} seconds is significant, because when we reach this extraordinarily early time the size of the visible universe becomes smaller than its quantum wavelength and is thus enshrouded by quantum uncertainty. When quantum uncertainty overtakes everything, we don't know the positions of anything, and we can't even determine the geometry of space. This is when Einstein's theory of gravitation breaks down.

---

*The amusing adventures of Mr. Tompkins, used by George Gamow to explain the ideas of physics to the lay reader, give a nice account of what the world would look like if the quantum wavelengths of things were close to their actual sizes. Playing pool becomes an unnerving experience for Mr. Tompkins.

This situation has inspired cosmologists to attempt to create a new theory of gravitation, in which gravity's quantum aspects are fully included, and to use it to find possible quantum universes. We are going to explore some of the ideas that have emerged from these bold investigations. They do not claim to be the final story—they may be only a tiny part of the final story—but the final story will doubtless be at least as radical in its treatment of our cherished cosmological notions.

In the pictures of possible beginnings to the expansion of the visible universe (figure 5.4), we illustrated what might happen to the size of the universe as it was traced into the past. In some options, there is an apparent beginning of time and space and everything else at a singularity. In others, space and time have always existed. But there is a more subtle possibility. Suppose that the very nature of time changed as it was traced back to the Planck time. The question of the beginning of the universe then becomes bound up with the issue of the nature of time itself.

....

CHAPTER 6

TIME—AN EVEN BRIEFER HISTORY

Brother Mycroft is coming round.
—*The Bruce-Partington Plans*

There is a longstanding puzzle about the true nature of time. One finds it confronted by thinkers in many cultures over thousands of years. It is this: should time be thought of as an unchanging and transcendent background stage upon which events are played out or just as the events themselves—so that if there were no happenings, there would be no time? The distinction is of interest to us, because the first assumption leads us to talk about the creation of the universe *in* time. The alternative is to think of time as something that comes into being along with the universe. There was no "before" the beginning of the universe, because once upon a time there was no time.

Our everyday experience of time measures it in terms of sequences of natural events: swings of a pendulum in the gravitational field of the Earth; the shadow cast by the sun on a sundial as the Earth rotates; or the vibrations of a cesium atom. We have no way of talking about what time "is" except in terms of how we measure it. Time is often defined by the way things change. If this were the correct way to do it, we could expect rather unusual things to happen to the nature of time as we encountered the extraordinary conditions that existed during the first moments after the big bang.

Isaac Newton's seventeenth-century picture of the world gave time a transcendental status. Time just passed, inexorably and uniformly, entirely unaffected by the events and contents of the universe. Einstein's picture of time was radically different. The geometry of space and the rate of flow of time were both determined by the material contents of the universe. Like the nature of Einstein's space, Einsteinian time is built upon his premise that no one has a privileged view of the universe. No matter where you are and how you are moving, you should deduce the same laws of physics from the experiments you perform.

This democratic treatment of observers in Einstein's general theory of relativity means that there is no preferred way of telling time in the universe. Nobody ever measures some absolute phenomenon called "time"; what one measures is the rate of some physical change in the universe. It could be the fall of sand in an egg-timer, the movement of the hands on a clock face, or the dripping of a tap. There are countless changing phenomena that could be used to define the passage of time. For instance, on a cosmic scale, observers around the universe could use the falling temperature of the background radiation to tell time. No one particular measure of change seems to be more fundamental than any other.

An illuminating way to view an entire universe of space and time—a "spacetime"—in Einstein's theory is as a pile of slices of space (imagine there to be only two dimensions of space rather than three, for the sake of visualization), with each slice in the stack representing the whole of space at a particular time. Time is just a label identifying each slice of space in the stack (as in figure 6.1). We can

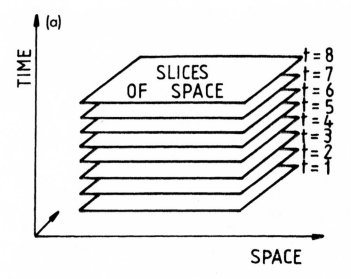

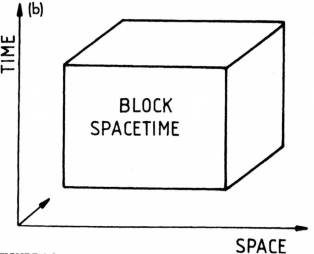

FIGURE 6.1
(a) A stack of slices of space taken at different times, here labeled $t = 1$ to $t = 8$; (b) a block of spacetime made from all the slices of space. This block could be sliced up in many ways that differ from the slicing chosen in (a).

see that the stack of spacetime can be sliced in many different ways—that is, at a variety of different angles. Each possible method of slicing will give us a different way of defining time. But the amalgam of spacetime itself is unaffected by the time slicing chosen, and so the spacetime stack is a more fundamental entity to focus upon than is either space or time separately.

In Einstein's description of space and time, the shape of the spacetime is determined by the matter and energy within it. This means that time can be defined by some geometrical property—like the curvature—of each slice, and hence in terms of the density and distribution of matter in the slice, since that is what determines its curvature. (Figure 6.2 gives a simple illustration.) Thus we begin to see a possibility of associating time—including its beginning and its end—with some property of the contents of the universe.

Despite having introduced these subtleties regarding the nature of time, general relativity fails to specify what the universe is like initially. Our spacetime stack always has a first slice, which determines what the others on top of it will look like.

In quantum theory, the nature of time is an even bigger mystery. If it is defined operationally, in terms of other properties of the universe, then it will suffer indirectly from the restrictions imposed upon our knowledge of these properties by quantum uncertainty. Any attempt to produce a quantum description of the universe will have very unusual consequences for our ideas about time. The most unusual has been the claim that a quantum cosmology permits us to describe a universe that has been created from nothing.

Simple cosmological models—ones that ignore the quantum nature of reality—can begin at a definite past

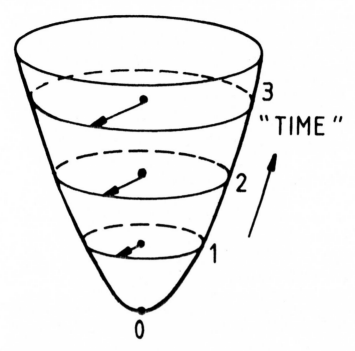

FIGURE 6.2
This dome-shaped spacetime is made up of a stack of circular disks of increasing radius. Each disk can be given a "time" labeled by the radius of the disk, so this geometrical "time" increases as we move up the dome from 0 to 3.

· · · · · · · · · ·

moment, which has been defined using certain types of clocks. The starting conditions that dictate the future behavior of the universe must be prescribed at that beginning. These models have been used to describe the present state of the universe, since the effect of quantum mechanics upon the universe today is tiny. But if we want to use those models near the Planck time, we need to understand how the description of time will be affected by the inclusion of quantum effects.

In quantum cosmology, time does not appear explicitly. It is a construct of the material contents of the universe and their configurations. Since we have equations that tell us something about how those configurations change as we look from one slice of space to another, it would be superfluous to have something called "time" in the story as well. The situation is not very different from that of a pendulum clock. The position of the hands on the clock face merely keep a record of how many swings the pendulum makes. There is no need to mention anything called "time," unless we want to. Likewise, in the cosmological setting, we distinguish the slices in our spacetime stack by the matter configuration shaping each slice. But this information about the distribution of matter is given to us only *statistically* by quantum theory. When we measure something, we find it to reside in any one of an infinite collection of possible states. Quantum mechanics tells us only the *probability* for it to be found in each of these states. The information that determines these probabilities is contained in a mathematical entity that has become known as the "wavefunction of the universe." We shall just call it W.

At present, cosmologists believe that they have a way to find the form of W. This might prove to be a blind alley, or even an outrageous oversimplification. More optimistically, we hope that it might at least be a pointer to a new and better approximation of the truth. The proposed route uses an equation first found by the American physicists John A. Wheeler and Bryce DeWitt. The Wheeler-DeWitt equation is an adaption of Erwin Schrödinger's famous equation governing the wavefunction of ordinary quantum mechanics, but with the curved-space attributes of general relativity incorporated into it. If we knew the present form of W, the equation would tell us the probability that the visible universe would be found to possess certain large-scale features. One hope is that the probability will turn out to be overwhelmingly great for particular large, expanding configurations of matter and radiation—in the same way that large everyday things have definite properties despite the minuscule uncertainties of quantum mechanics. If the most likely values do indeed correspond to situations observed by astronomers (for example, by predicting certain patterns of galaxy clustering or certain temperature variations in the microwave background radiation), then many cosmologists would be satisfied that ours was one of the most "probable" of all possible universes. However, in order to use the Wheeler-DeWitt equation to find W for the cool, low-density universe that we observe now, we need to know what W was when the universe was at its maximum density and temperature—that is, when it "began."

The most useful quantity involved in the manipulation and study of W is the transition function, which tells us the likelihood of particular changes occurring in the state of the universe. We denote it by T, so that $T[x_1,t_1 \rightarrow x_2,t_2]$ gives the probability of finding the universe in a state x_2 at a time t_2 if it was in a state x_1 at an earlier time t_1, where the "times" are specified by some attribute of the state of the universe—for example, its average density.

In nonquantum physics, the laws of nature dictate that a particular future state will arise from a particular past one; we do not talk about probabilities. But in quantum physics—as the American physicist Richard Feynman taught us—a future state is determined only by some appropriate average of all the possible paths through space and time that history could have taken. One of these paths might be the unique one dictated by the nonquantum laws of nature. We call this the "classical path." In some situations, the quantum description has a transition function

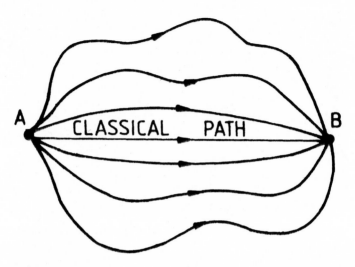

FIGURE 6.3
Possible paths between A and B: the Newtonian laws of motion dictate that the "classical path" is taken. Quantum mechanics gives a probability for the transition from A to B which is an average over all the possible paths between A and B, some of which are indicated here.

· · · · · · · · · ·

that is largely determined by the classical path—the other paths having combined to cancel each other out, rather like the peaks and troughs of waves that are out of phase (see figure 6.3).

It is a deep question whether all possible starting states for a quantum universe with very high density can give rise to a universe like our own. Ours is a universe in which quantum uncertainties are small, and there is an unambiguous feeling of the flow of "time" in everyday experience. The requirements for a universe like ours—one that permits the existence of living observers—may turn out to be very restrictive, marking our universe as special among all possible worlds.

In practice, W depends upon the configuration of all of the matter and energy in the universe on a particular slice through the spacetime stack, and upon some intrinsic aspect of the slice (like its curvature) that effectively labels its "time" uniquely by identifying the slice in the stack. The Wheeler-DeWitt equation then tells you how the wavefunction for one value of this intrinsic time is related to its form at another value of intrinsic time. When we are near the classical path, these developments of the wavefunction are straightforward to interpret as small modifications to ordinary classical physics. But when the most probable path is far from the classical one, it becomes increasingly difficult to interpret the quantum evolution as occurring "in" time in any sense—that is, the collection of space slices that the Wheeler-DeWitt equation gives us do not stack up to look like a spacetime. Nonetheless, the transition functions telling us the probabilities for the universe to pass from one state to another can still be found. The question of the starting state of the wavefunction now becomes the quantum analog of the search for the origin of the universe.

The transition function tells us the probability of the universe making a transition from one geometrical configuration of matter to another. This development from one configuration to another is shown in figure 6.4.

We can envisage universes that begin as a single point rather than as an initial space slice. They look conical rather than cylindrical (as is the case in figure 6.4). This is illustrated in figure 6.5.

Yet this is no real advance, because any singularity in the nonquantum cosmological models will show up as a singular feature of the classical path, and we just seem to be picking a particular initial condition—which happens to describe creation from an initial preexistent point—for no good reason.

There is a radical step that may now be taken. We should stress that it may well turn out to be empty of any physical significance. It is an article of faith, guided by esthetics. Look at figures 6.4 and 6.5 and notice how the stipulation of an initial condition g_1 relates to the state of the space further up the tube (or the cone) at g_2. Perhaps the boundaries of the configurations at g_1 and g_2 could be combined in some way so that they describe a single smooth space, like that in figure 6.6, which contains no nasty singularities.

We know of simple two-dimensional examples, like the surface of a ball, that are smooth and free of any singular points like the one that occurs at the point of a cone. So we could think of the whole boundary of the four-dimensional

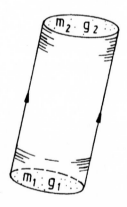

FIGURE 6.4

Some spacetime paths whose boundary is composed of two three-dimensional spaces with curvatures labeled g_1 and g_2 and containing material in some arrangements m_1 and m_2, respectively. The boundary regions are shown shaded and are drawn here as two-dimensional ends of a three-dimensional cylinder.

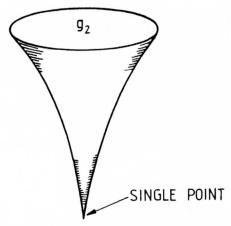

FIGURE 6.5
A spacetime path with a boundary composed of a curved three-dimensional space g_2 and a single initial point.

• • • • • • • • • •

spacetime as being not g_1 and g_2 but a single smooth surface in three dimensions. It would be like the surface of a ball that exists in four spatial dimensions. The surfaces of balls have the interesting feature of being finite in size but having no edge: such a surface has a finite area (it would require only a finite amount of paint to paint it), but when you move around on it you can never run off the edge or encounter a sharp point, like the apex of a cone. The surface of a ball has *no* boundary, as far as its inhabitants are concerned. An analogous situation can be imagined for the initial state of the universe. However—and now comes the radical step—the ball we have been using as an example occupies a space of three dimensions and has a two-dimensional surface. For our quantum geometry we need a three-dimensional surface, in four-dimensional *space* (*not* four-dimensional *spacetime*, which is what the real uni-

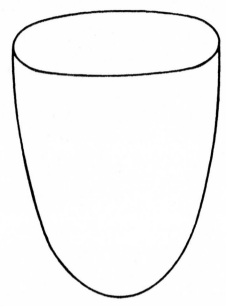

FIGURE 6.6

A path whose boundary has been smoothly rounded off so that it consists of a single three-dimensional space, with no point at the base as was the case in figure 6.5. This allows its transition probability to be interpreted as that for a universe that has been created out of nothing at all.

• • • • • • • • •

verse is assumed to be). Therefore, in 1983, it was proposed by Stephen Hawking and the American physicist James Hartle that our ordinary concept of time is transcended in this quantum-cosmological setting and becomes just another dimension of space.

This is not quite as mystical as it sounds, because physicists have often used this change-time-into-space trick for solving certain problems in ordinary quantum mechanics, although they do not imagine that time *really* becomes space. At the end of the calculation, they simply revert to the usual interpretation of there being one dimension of time and three (qualitatively different) dimensions of space. It's just like using another language temporarily.

Let us digress for a moment. One of the most interesting things about this time-becomes-space idea is how difficult it is to convey in words a good picture of what is going on. Hawking's 1988 book *A Brief History of Time* was the first attempt to do this. The popularization of science has come to mean the explanation of complicated mathematical abstractions in terms of simple, visualizable pictures or analogies. Writers often liken interactions between elementary particles to collisions between billiard balls, or describe atoms as mini solar systems, and so forth. In fact, at the end of the nineteenth century some French mathematicians were rather critical of physicists who insisted on a mechanical picture of various physical phenomena, complete with little rolling balls, wheels, and pieces of string. What the popularizers are taking advantage of is the fact that there exist simple analogies between the more esoteric workings of the universe and other events in our experience. But the idea that time becomes another dimension of space does not seem to have a nice familiar analogy. One can read the sentence "Time becomes another dimension of space," understand what all the words mean, but still not possess any real understanding of what is being conveyed. This lack of a ready analogy may have been a source of difficulty for readers of *A Brief History of Time*. We have come to expect that the deepest facets of the makeup of the universe—the depths of the inner space of elementary particles or the outer space of galaxies and black holes—will have simple local analogies. This may not be so. Indeed, the absence of analogies may be a good sign that we are touching upon some brute fact of reality rather than just redeploying our old familiar concepts.

The radical character of this quantum approach to time

is that it treats time as being truly *like* space, in the ultimate quantum-gravitational environment of the big bang. As one moves away from the beginning of the universe, one expects that quantum effects should start to interfere with each other, as wave crests meet wave troughs, and that the universe will follow the classical path with greater and greater accuracy. The conventional character of time, as qualitatively distinct from space, begins to crystallize out in the first moments after the Planck time. Conversely, as one goes back toward the beginning, the distinct character of time melts away and time becomes indistinguishable from space.

This timelessness for the original quantum state of the universe was proposed by Hartle and Hawking because it looked so economical and because it avoids a singularity in the starting state. For these reasons, it has become known as "the no-boundary condition." More precisely, the no-boundary proposal stipulates that the wavefunction of the universe is determined by an average of transitions that are restricted to four-dimensional spaces with a single, finite, smooth boundary, like the spherical one we discussed earlier.

The transition probability that this recipe provides has a form in which there are no prior initial states. Hence, the no-boundary condition is often described as calling for "creation out of nothing," because it is a picture in which T gives the probability for a certain type of universe to have been created out of nothing. As a consequence of the "time-becomes-space" proposal, there is no definite moment or point of creation.

The overall picture one gets of this quantum beginning is that as one looks back toward that instant we have called the "zero" of time, the very notion of time fades away and ceases to exist. This type of quantum universe has not always existed—it comes into being, just as nonquantum cosmologies with singularities seem to—but it does not start with a big bang, where physical quantities are infinite and further initial conditions need to be specified. In neither the singular big-bang creation nor the quantum creation is there any information as to what the universe may have come into being from, or why.

We should stress again that the Hartle-Hawking proposal is a radical proposal. It has two ingredients: the first is that "time becomes space"; the second is the no-boundary condition. It is a prescription for the state of the universe which subsumes the roles of the initial conditions and the laws of nature in the traditional picture. Even if one subscribes to the first ingredient, there are many choices one could have used instead of the no-boundary condition to specify the state of a universe that tunnels into existence out of nothing.

In figure 6.7 is shown the variation of the wavefunction of the universe W with the density of the universe (the "clock"), in the case where the no-boundary condition is used and also for another possible boundary condition, suggested by the American physicist Alex Vilenkin, which has a quite different character. Large values of W correspond to high probabilities. So, we see that with the no-boundary condition it is most improbable that the universe came into existence with a high density, while the Vilenkin condition makes this very probable. Some critics of the no-boundary condition feel that it is unlikely to lead to a very early universe dense enough and hot enough to undergo inflation.

The study of the wavefunction of the universe is in its infancy. Ideas about it will undoubtedly change in many ways before the theory is complete. The no-boundary condition leaves something to be desired. It does not specify

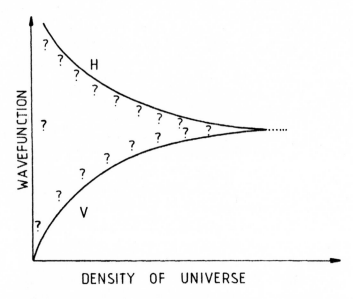

FIGURE 6.7

The possible variations of the wavefunction of the universe with the density of matter in the universe. A high value of the wavefunction corresponds to a high probability of occurrence. The Hartle-Hawking (H) and Vilenkin (V) proposals for the wavefunction are shown. A variety of other possible behaviors might be allowed in between (the "?" region). The theory becomes very unreliable at very high density (shown dotted).

··········

the small nonuniformities that are needed if galaxies are to form. It must be supplemented by additional information about the matter fields in the universe and how they distribute themselves. It may be right; it may be half-right; it may be wrong. The pessimist might even argue that we will never be able to tell, because the universe may be fashioned in a manner that leaves no traces of its quantum origins—or none significant enough for us to observe today and so test our ideas against the facts. This could be the case if inflation occurred.

The important lesson for us to draw is the extent to which our traditional way of thinking about the evolution of the universe—in terms of starting conditions acted upon by laws of change—could be mistaken. It might be an artifact of our experience of a realm of nature in which quantum-gravitational effects are vanishingly small. The no-boundary condition and its various rivals appear to have been selected partly for reasons of their simplicity, or the ease with which they allow calculations to be done. They are not, as far as we know, demanded by the internal logic of the quantum universe.

The view that starting conditions are independent of the laws of nature must be reassessed in the case of the initial state of the universe. If the universe is unique—because it is the only logically consistent possibility—then the initial conditions are also unique and become a law of nature themselves. On the other hand, if we believe that there are many possible universes—indeed, that there may actually *be* many other universes—then initial conditions need have no special status. They could all be realized, somewhere.

The traditional view that initial conditions are for the theologians and the laws of change for the physicists seems to have been clouded—at least temporarily. Cosmologists now engage in the study of initial conditions to discover whether there exists a "law" of initial conditions, of which the no-boundary proposal is just one possible example. This proposal is certainly radical, but perhaps it is not radical enough. It is worrying that so many of the concepts of the modern quantum-cosmological picture—"creation out of nothing," "time coming into being with the universe"—are just refined images of traditional human intuitions and categories of thought that medieval theologians would have felt quite at home with. Surely, these traditional notions have suggested many of the modern cosmological concepts, even though these concepts are cast in mathematical form. The time-becomes-space proposal of Hartle and Hawking is the one truly radical element of cosmology that we cannot detect as the legacy of past generations of thinking in philosophy or theology. One suspects that a good many habitual concepts may need to be discarded before the true picture emerges.

Despite the confidence with which some modern cosmologists have addressed questions about the origin of the universe—a confidence that has seen the publication of research papers bearing titles like "The Creation of the Universe out of Nothing"—one should be cautious. All these theories need to assume at the outset the existence of a good deal more than one's everyday conception of "nothing" in order to say anything of interest. In the beginning, there must exist laws of nature (the Wheeler-DeWitt equation, in our discussion), energy, mass, geometry; and, of course, underpinning everything seems to be the ubiquitous world of mathematics and logic. There needs to be a considerable substructure of rationality before any complete explanation for the universe can be erected and sustained. It is this underlying rationality that most modern theologians emphasize when questioned about the role of God in the universe. They do not regard the Deity as simply the Great Initiator of the expansion of the universe.

That part of the scientific enterprise that tries to explain the existence of the universe as a consequence of a prior state consisting of *absolutely nothing* violates our deep-rooted sense that "there is no such thing as a free lunch." Nonscientists take it for granted that you can't manufacture something out of nothing. If one proposes to give a scientific account of a universe coming into being, an immediate objection seems to be that one would indeed be trying to get something out of nothing, because one would have to bring into being a universe that possessed energy, angular momentum, and electric charge. This would violate the laws of nature, which enshrine the conservation of these quantities, and so the creation of the universe out of nothing cannot be a consequence of those laws.

This argument is really quite persuasive, until one starts to inquire what the energy, angular momentum, and electric charge of the universe might be. If the universe itself possesses angular momentum, then on the largest scale the expansion will possess rotation. The most distant galaxies would be moving *across* the sky as well as receding from us. Although this lateral motion would be too slow for us to observe, there are other sensitive indicators of any cosmic rotation. If we consider the effects of the Earth's rotation, we see that it causes a slight flattening of the Earth at the poles. A similar phenomenon would occur if the universe were rotating: directions along the rotation axis would expand more slowly than others. Consequently, the microwave background radiation would be hottest if it came from the direction of the rotation axis and coolest from directions at right angles to that. The fact that the radiation temperature is the same in every direction to a precision of one part in one hundred thousand means that if the universe does rotate it must be rotating more than one trillion times more slowly than it is expanding in size. This ratio is tiny enough to suggest that the universe may have *zero* net rotation and angular momentum.

Similarly, there is no evidence that the universe possesses any overall net electric charge. If any cosmic structures possess an electric charge—because of an imbalance between, say, the numbers of protons and electrons within them—this imbalance would have a dramatic effect upon the expansion of the universe, since electricity is so much stronger than the force of gravity. In fact, a remarkable consequence of Einstein's theory of gravitation is that a "closed" universe—one that will contract to a future singularity—*must* have zero total electric charge; that is, the individual electric charges of all the matter it contains must total to give zero overall charge.

Finally, what about the energy of the universe? This is the most intuitively familiar example of something that you cannot produce from nothing. But, remarkably, if the universe is closed it must also have *zero* total energy. The reason can be traced to Einstein's formula $E = mc^2$, which reminds us that mass and energy are interchangeable and that we should be thinking about the conservation of *mass-energy* rather than of energy or mass alone. The important point is that energy, in forms other than mass, comes in positive and negative varieties. If we add up all the masses in a closed universe, they contribute a large positive contribution to the total mass energy. But those masses also exert gravitational force upon each other. This force is equivalent to negative energy—or what we call "potential energy." If we hold a ball in our hand, it has potential energy of this sort: when the ball is dropped to the ground, a positive energy of motion is created at its expense. The law of gravitation insures that the negative potential energy of gravitation between the masses in the universe must always be equal in magnitude but opposite in effect to the sum of the mc^2 energies associated with each of the individual masses. The total is therefore always exactly zero!

Here is a remarkable state of affairs. It seems that universal values of the three conserved quantities that prevent us getting something from nothing may well all be equal to zero. The full implications of this are still not clear. But it seems that nature's conservation laws need not present a barrier to the appearance of a universe out of nothing (or, for that matter, to its disappearance back into nothing). The laws of nature may well be able to describe the creation process.

To close this discussion of the scientific ramifications of creation out of nothing, we should hark back to the earlier idea that the universe began at a singularity of space and time. The no-boundary condition of quantum cosmology avoids the necessity for such a cataclysmic beginning, and is therefore currently fashionable among cosmologists. However, one should be wary of the fact that many of the studies of quantum cosmology are motivated by the desire to avoid an initial singularity of infinite density, so they tend to focus on quantum cosmologies that avoid a singularity at the expense of those that might contain one. It is worth noticing that the traditional big-bang picture of the universe emerging from a singularity is, strictly speaking, also creation out of absolutely nothing. No cause is given, and no restrictions are placed upon the form of universe that appears. There is no prior time, no prior space, and no prior matter. Researchers into quantum creation hope that by pursuing a description of a highly probable universe from some inevitable quantum state we will learn why our universe possesses so many unusual properties. Unfortunately, many of those properties could have arisen from a later period of inflationary expansion, and inflation can arise from a wide range of initial quantum states.

A long shot, Watson; a very long shot!
—*Silver Blaze*

All the things around us, from cabbages to kings, have the density and the hardness they do because of certain unchanging aspects of the fabric of the universe. These unchanging aspects are called the "constants of nature." They are fixed values for such phenomena as the strength of the force of gravity, the masses of the elementary particles of matter, the strength of electricity and magnetism, and the speed of light in empty space. They are called "fundamental" constants if they cannot be expressed in terms of other constants of nature. We are able to measure most of these quantities with great precision. Their numerical values are what distinguish our universe from others we might imagine which obey the same physical laws. However, although these constant quantities appear in all our laws of nature, they are at root the deepest mystery about the structure of the universe. *Why* do they have the particular values they do? It has always been the dream of physicists to come up with some complete theory of physics in which the values of the fundamental constants are predicted or explained. Many great scientists have tried: all have failed to make any headway with this problem.

The recent attempts to develop a quantum description of the universe and its initial state unexpectedly yielded a possible way to explain the values of the constants of nature. The general idea of the search for the wavefunction of the universe, initiated by James Hartle and Stephen Hawking, was to suppose that the universe, at those extreme densities where its quantum attributes become overwhelming, behaves like a four-dimensional ball. But then some cosmologists began to ask what would happen if the surface of the ball were not uniformly smooth: suppose there were tubes joining one part of the surface to another (see figure 7.1). These tubular connections came to be called "wormholes." They are connections between regions of spacetime that would otherwise be inaccessible to each other.

This elaboration has arisen for several reasons. One is the tendency of physicists to tinker with their picture of the world in order to discover whether something new will emerge to provide an explanation for one or another unsolved puzzle of nature. But there is a more specific imperative. The intuitive picture that existed of the state of spacetime at and before the Planck time, 10^{-43} seconds, was that of a turbulent foam dominated by quantum uncertainty. The presence of wormholes with a diameter equal to the distance light has traveled by this time (about 10^{-33} centimeters) is a likely consequence of the chaotically interconnected state of space.

This enlargement of our picture of the global nature of space produces a mind-boggling increase in the possible complexity of the universe. It could consist of a large number (or even an infinite number) of extended regions of space connected to themselves and to each other by worm-

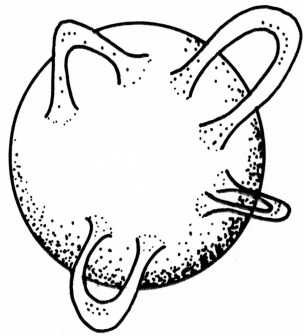

FIGURE 7.1
A space with wormhole connections to itself.

. . . .

holes. In figure 7.2, we depict a situation in which there are a number of interconnected "baby universes."

In order to make sense of what is occurring in situations like this, let us consider only the simplest type of wormhole connections, in which wormholes are allowed to connect only the baby universes. This simplification is called the "dilute wormhole approximation," because it is analogous to a simplifying assumption used to describe the behavior of ordinary gases. The dilute gas approximation can be made because gas molecules spend much longer traveling in between collisions than they do in the

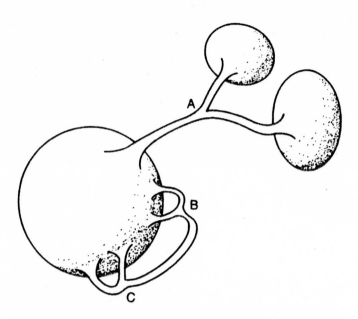

FIGURE 7.2
A network of wormholes that does not adhere to the dilute
wormhole approximation: wormholes split from the "mother
universe" to form two "baby universes" (at A) and join
wormholes to other wormholes (at B and C) on the mother
universe.

· · · · · · · · · ·

collision process. When this is not true—in cases, say,
where the gas is condensing into a liquid—the behavior is
much more interactive. The dilute wormhole approximation is a simplification of the type of interactions permit-
ted between the baby universes. It assumes that the wormholes connect only large smooth regions and that they do
not split into two tubes or join up with other wormholes
(see figure 7.3).

This would all be very pretty, but not much else, if it
were simply what it appears to be—a generalization for
generalization's sake. But the wormhole scheme turned

out to offer much more. The values of the constants of nature that exist in any large region of the universe might now be determined by the network of fluctuating wormhole connections to that region. But because the wormhole connections possess all the attributes of quantum uncertainty, the constants will not be determined exactly but only statistically.

The simplest constant to study was the famous "cosmological constant"—the term Einstein introduced into his equations of general relativity to produce a static model of

FIGURE 7.3

A number of "baby universes" connected by wormholes and possessing other wormhole connections to themselves. These wormholes do not join wormholes to other wormholes, nor do they split into two or more wormholes. This state of affairs is called the "dilute wormhole approximation."

the universe, and later discarded. The cosmological constant created a long-range force of repulsion to oppose the attractive force of gravity between masses. Although one could simply ignore the possibility of this addition to the law of gravitation, as cosmologists generally did, there is no known reason why it should not appear in Einstein's equations. It's a nasty loose end. Even if it cannot stop the universe from expanding, it could still change the rate at which the universe expands today. Astronomical observations of the rate of expansion of the universe show that the cosmological constant, if it exists, is amazingly small. Expressed as a pure number, it must be smaller than 10^{-120}! This number is so small that it suggests there might well be some unknown law of nature which demands that it actually be *zero*. However, quite the opposite circumstance has emerged from all the studies of the behavior of the elementary particles and energy fields existing in the early universe. Not only do these studies predict that a cosmological constant is to be expected: they predict that its size should be huge—vastly bigger than allowed by our observations of the expansion today—perhaps even 10^{120} times bigger!

In 1988, the American physicist Sidney Coleman made a remarkable discovery. If a universe were to begin with a cosmological constant in addition to gravity, the effect that it would have upon wormholes would be to create an opposing stress that would cancel its own antigravitational effect up to the level of intrinsic quantum certainty. Inclusion of the wormhole fluctuations therefore leads to the prediction that when a baby universe becomes large (like our visible universe today) the most overwhelmingly probable value of the cosmological constant within it is zero, as shown in figure 7.4.

So far, this success has not been extended to derive a prediction for one of the nonzero constants of nature, like

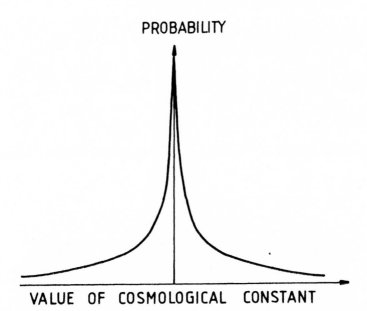

PROBABILITY

VALUE OF COSMOLOGICAL CONSTANT

FIGURE 7.4
The probability of the cosmological constant having a particular
value as a result of wormhole fluctuations. The most likely
value is very sharply peaked around zero.

· · · · · · · · · ·

the mass or the electric charge of the electron. However, it
is illuminating to consider the possible nature and interpretation of such a prediction.

Suppose we were able to calculate a probability spread
for a fundamental constant—say, the strength of the electromagnetic force—in the universe today. The result might
look like any one of those shown in figure 7.5.

In the first case, all values of the constant are equally
likely, and the wormhole theory does not make a prediction that we can check against the observed value of the
constant. In the second case, the constant is overwhelmingly likely to have a value at the peak of the graph. Most cosmologists interpret such a peak as picking out the situation

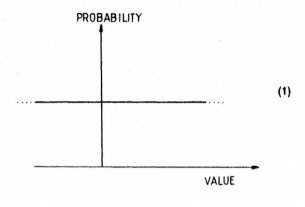

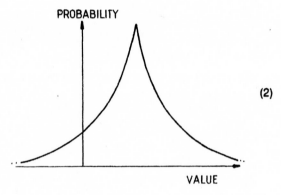

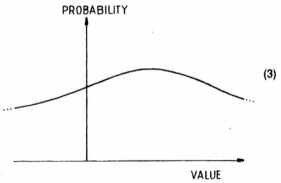

FIGURE 7.5
Three possible predictions for the observed values of constants of
nature which could emerge from a wormhole theory: (1) equal
likelihood of any value; (2) one value clearly more likely; (3)
likelihood spread over many values, with no very pronounced peak.

we should observe, because it identifies the most probable one. If the probability distribution for the expected value of Newton's constant of gravitation was strongly peaked around the observed value, we would regard this as an amazing success for the wormhole theory. It would also allow us to use observations of the constants of nature to probe our theories of quantum gravity before the Planck time. Unfortunately, it has proved too difficult to extract such predictions from the theory.

As we have seen, many physicists believe that there must exist a single description of the laws of nature which unites all we know about the disparate forces of gravity, electricity, magnetism, radioactivity, and nuclear physics. This unified expression of nature's laws has been dubbed the Theory of Everything, and one of the hopes physicists have for it is that it will require the constants of nature to have one and only one set of logically consistent values. If we find the Theory of Everything, it should tell us the values of the fundamental constants—and this would be the ultimate test of such a theory. However, even if a Theory of Everything fixed the initial values of the constants of nature in each "baby" and "mother" universe, the wormhole connections between them would produce unpredictable fluctuations that would shift the values of those constants. Their measured values would drift away from those they were given *ab initio*. Consequently, their observed values today need not tally with those fixed by the Theory of Everything.

Let us now consider the last of the three hypothetical cases in figure 7.5. In (3), the probability is spread fairly evenly over a wide range of possible values. There is a most probable value—but only just. This raises all sorts of awkward questions. Why should we be comparing the observations of our universe with the predictions for the most probable universe? Should we expect our universe to be among the "most probable," in some quantum sense? We shall argue that there is every reason to expect that our universe is *not* among the most probable.

In the opening chapter of our story, we introduced the concept of an expanding universe and showed how the age of such a universe is strongly linked to the evolution of observers. An old universe is necessary to produce stars, which generate the nuclear elements heavier than helium required for the subsequent evolution of complexity. Similarly, we can consider why the existence of observers like ourselves (or even observers unlike ourselves) means that the constants of nature must have numerical values not too unlike those we observe. Were the strength of gravity a little different, or the strength of the electromagnetic force slightly perturbed, then stable stars could not exist and the finely balanced life-enabling properties of nuclei, atoms, and molecules would be destroyed. Biologists believe that the spontaneous evolution of life requires the presence of carbon, with all the bonding properties that make it the basis of DNA and RNA, the helical molecules of life. The presence of carbon in the universe depends not merely upon the age and size of the universe but also upon two amazing apparent coincidences between those constants of nature that determine the energy levels of nuclei. When nuclear reactions in the stars combine two helium nuclei to produce beryllium, we are just one step away from making carbon by the addition of another helium nucleus. But this reaction appears to be too slow to make carbon of any consequential amount in the universe. Prodded by the fact that we do indeed exist, Fred Hoyle made a startling prediction back in 1952. He predicted that the carbon nucleus could reside in an energy level just greater than the sum of the energies of the helium and beryllium nuclei. This situation produces an especially fast helium-beryllium reaction, because the combination of the two nuclei occupies what is called a "resonant" state: one that has a natural level of energy waiting for it. Hoyle turned out to be right. Nuclear physicists were amazed to find a previously unknown energy level of the carbon nucleus exactly where he had predicted it would be. The Caltech physicist William Fowler, who won the Nobel Prize in recognition of his immense contributions to the field of nuclear astrophysics, once remarked that it was Hoyle's prediction that convinced him he ought to work in this field. If someone could tell him where to find a nuclear energy level just by thinking about the stars, there had to be something in this astrophysics business after all!

If the constants of nature were slightly different, the resonance of helium, beryllium, and carbon would not exist—and neither would we, because there would be hardly any carbon in the universe. And here is the second coincidence: Once the carbon is made, it could all be turned into oxygen by the nuclear reactions between carbon and further helium nuclei. But this reaction just fails to be resonant—by an even finer margin—and therefore the carbon survives.

What these examples teach us is that the existence of complex structures in the universe is made possible by a combination of apparent coincidences regarding the values of the constants of nature. Were those values to be slightly changed, the possibility of conscious observers evolving would disappear. We cannot draw any grand philosophical or theological conclusions from this fortunate state of affairs. We cannot say that the universe was "designed" with living observers in mind, or that life had to exist, or that it exists elsewhere in the universe, or even that it will continue to exist. Any or all of these conjectures might be either true or false. We simply have no way of telling at the moment. All we need to appreciate is that in order for a universe to contain living observers (or even just atoms or their nuclei) the constants of nature—or, at any rate, a very large number of them—need to have values very close to those observed.

With this in mind, we consider figure 7.5(3) again. Mark the narrow range of values of the constant which permits the subsequent evolution of biological complexity, and reconsider. The range permitting observers will be very small and may lie far from the most probable value that a theory predicts (see figure 7.6). Comparing the theory with observation now becomes very difficult. We are not really

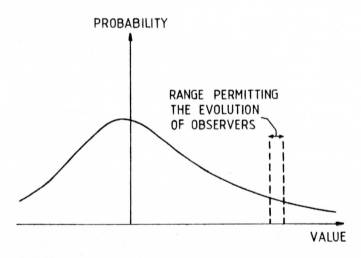

FIGURE 7.6
A possible prediction for the probability of finding a constant having a particular value in the universe today. The range of values that permit the evolution of "observers" is also indicated. This range appears to be very narrow for most of the fundamental constants of nature. It may also lie far from the most probable values of the constant, as we have displayed it here.

interested in the most probable values of the constants. We should be interested only in the most probable values that permit the evolution of observers. To exaggerate, if the most probable value of the strength of gravity leads to universes that live for a billionth of a second, then we cannot be living in the most probable universe.

We have learned a very important lesson. When we have a cosmological theory that makes statistical predictions about the structure of a universe that emerges from quantum origins, then in order to test those predictions against the observed facts we have to know *any and every way* in which the predicted quantity is necessary for the evolution of observers. The range of life-permitting values of that quantity might be very small and extremely improbable from an absolute point of view. Nonetheless, we are obliged to reside in such an improbable universe, because we could exist in no other. Our tortuous journey through the labyrinth of wormholes to the beginnings of time has brought us back four-square to the fact of our own existence as an important datum in our search for the origins of the universe and its remarkable panoply of properties.

The only escape from these conclusions is to suppose that "life" is a generic phenomenon—one that would arise by hook or by crook, regardless of the values of the constants of nature. This is hard to reconcile with our knowledge and experience of life. The evolution of conscious life in particular (rather than just complicated molecules) seems a fairly precarious business, even with the values of the constants that we do have. Biologists emphasize the enormous number of evolutionary pathways that lead to dead ends. We don't deny the possibility that there are lots of other life-forms around the universe today, but we believe that they must be atom-based—and, indeed, carbon-based—if they have evolved spontaneously.

Other types of life can certainly exist; for example, we are on the way to producing simple forms of silicon-based life. At present, the study of what has become known as "artificial life" (as opposed to "artificial intelligence") is a fascinating growth area of science. It brings together physicists, chemists, mathematicians, biologists, and computer scientists to study the properties of emergent complex systems that possess some or all of the properties we associate with "living" things. Most of these studies exploit fast computer graphics to simulate the behavior of complex systems interacting with their environment, growing, replicating, and so on. Whether this can really be called "living" remains to be seen, but ultimately such studies should shed important light upon the conditions essential for the emergence of structures complex enough to be called "conscious observers."

How often have I said to you that when you have
eliminated the impossible, whatever remains,
however improbable, must be the truth?
—*The Sign of Four*

Since the mid-1980s, the search for a Theory of Everything
has been dominated by the concept of superstrings. Whereas earlier searches for the ultimate laws of particle physics
focused upon mathematical descriptions in which the
most basic entities were points with no size, superstring
theory uses lines, or loops, of energy as the most basic
ingredients. The "super" prefix refers to the special symmetry that these strings possess which enables them to
bring about a unification in the description of the elementary particles of matter and the different forms of radiation
in nature. The idea of the most elementary particles being
like little loops seems peculiar, but these loops are more
like elastic bands: they possess a tension that depends
upon the temperature of the environment. At low temperature, the tension is very high and the loops will contract to
behave like points. Hence in the relatively moderate conditions that prevail in the universe today, strings have pointlike behavior to high precision, and they will fit the pre-

dictions of low-energy physics as do the pointlike elementary particles. However, it has long been known that the pointlike picture produces nonsensical results when applied to conditions of very high energy or temperature. Moreover, the pointlike picture adamantly refuses to allow us to bring gravity into harmony with the three other forces—electromagnetism and the strong and weak nuclear forces. In contrast, the string theory behaves beautifully at high temperatures, and gravity—far from being excluded— is required to join hands with the other forces of nature. The nonsensical answers disappear, and all the observable properties of elementary-particle physics can in principle be computed from the theory (although no one has yet been smart enough to do this).

All this sounds marvelous. But there is a snag. Superstring theories may have these much-sought-after properties only if they live in universes that have many more dimensions of space than the three we are familiar with. The first models that were constructed required either nine or twenty-five dimensions of space! The search then began for a natural process, occurring close to the Planck epoch, which would insure that if the universe began with, say, nine space dimensions all expanding equally, six of those dimensions would remain trapped at the size of the universe at that time—10^{-33} centimeters—while the remaining three continued to expand until they are now 10^{60} times as big as the rest (see figure 8.1). Today, according to this theory, the extra dimensions remain trapped at the scale of Planck length, so their effects are indiscernible—not just in everyday experience but also in the events created so far in high-energy physics experiments.

How this trapping might occur is still an unsolved problem. If it did occur, it makes the study of the very early universe much more difficult. It could be that

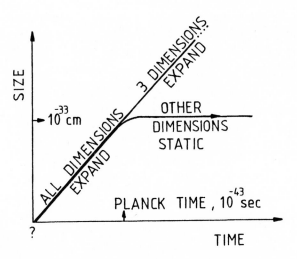

FIGURE 8.1

The proposed variation with time of the sizes to which the various dimensions of space expand in a conceivable superstring universe. It begins with all dimensions expanding in the same way, but after the Planck time, 10^{-43} seconds, only three of the dimensions of space continue to expand, becoming those we experience today. They are at least 10^{27} centimeters in size and form the space of the visible universe. The rest stay trapped and static; they would be imperceptible to us now, because they inhabit a universe only 10^{-33} centimeters in extent. As yet, there is no observational evidence that these additional dimensions of space exist.

....

there is some deep principle of nature requiring that three and only three dimensions of space continue expanding and become very large, like those we experience in the universe today. Alternatively, the number of large dimensions could be determined quite randomly. That number might even be different from one region of the universe to the next.

The number of large dimensions to space is something that plays a key role in what can occur in the universe.

Remarkably, universes with three large dimensions of space are very special. If there are more than three large dimensions, no stable atoms can exist, nor can there be any stable planetary orbits around stars. Waves also behave in a unique fashion in three dimensions. If the number of dimensions of space is even—say, two, or four, or six—then wave signals reverberate; that is, wave signals sent out at different times can arrive at the same time. In odd-numbered dimensions, this does not occur; wave signals are reverberation-free. However, in all the odd-numbered dimensions other than three, wave signals will be distorted. Only in three dimensions do waves propagate in a sharp, undistorted fashion. For these reasons, it appears that living observers can exist only in universes with three large dimensions (although there have been interesting speculations about what might be possible in two dimensions), because of the absence in additional large dimensions of any structures (like atoms) bound together by electromagnetism and the strong nuclear force.

We see that if there are three large dimensions of space because of some deep principle of nature, then we are very fortunate. If the number of the world's dimensions is a random outcome of events near the beginning of time—or varies from place to place beyond the horizon of the visible universe today—then the situation is rather like that with the determination of the constants of nature by wormhole fluctuations. We might determine the *probability* that we will find three dimensions of space, but no matter how small that probability turns out to be we know that we might have to find ourselves observing a universe with precisely three large dimensions of space, because we could have evolved in no other.

The speculative directions in which the frontiers of cosmology and high-energy physics are pushing as they explore the ramifications of new mathematical theories have highlighted one general feature of cosmology. It does not fit neatly into traditional attempts to define science. Philosophers of science, such as Karl Popper, stress the necessity for statements to be testable in some way if they are to be meaningful or "scientific." In laboratory-based sciences, this creates few problems. One can carry out virtually any experiment one chooses, in principle—although in practice there might be financial, legal, or moral constraints upon what one can do. In astronomy, the situation is different. We are not at liberty to carry out experiments on the universe; we can observe it in a variety of ways, but we cannot experiment directly upon it. Instead of conducting experiments, we search for correlations between things. If we observe many galaxies, we note whether all the very large ones are also very luminous, whether the spiral-shaped ones contain the most gas and dust, and so on. In cosmology, too, the situation differs from that of terrestrial sciences, in that our observations of the universe are biased in a way that cannot be corrected for simply by repeating the experiment under different conditions. We have explained why we necessarily live after the universe has been expanding for billions of years, and why we can see only a fraction of the entire (possibly infinite) universe. We have also noted that one consequence of the properties of the universe varying from place to place is that observers can evolve only in particular regions. Cosmology is a study in which the available data will always fall short of what one would like. Moreover, some of our data are biased in another way. Bright galaxies are easier to see than faint ones. Optical light is easier to detect than X rays. The art of being a good observational astronomer is to understand the biases that the data-gathering process might be introducing into your observations.

Bearing in mind these characteristics of cosmology, it is interesting to look at a growing trend in studies of the origin of the universe. We stressed earlier the contrast between those who seek to explain the observed structure of the universe in terms of what it was like when it began and those who try to show that its present structure is the inevitable result of past physical processes regardless of how it began. The inflationary-universe picture is the fullest manifestation of the second approach. No matter how the universe began, it is argued, there would have been some region, small enough to be kept smooth by interactions between matter and radiation, that could have undergone a period of accelerated expansion. The result is a universe that looks very much like our own: old, big, containing no magnetic monopoles, and expanding tantalizingly close to the critical divide separating "open" from "closed" universes. But in recent years there has been a focus upon the first approach as well. Scientists have begun to investigate whether there are principles that dictate the initial state of the universe. In effect, one is asking for a new sort of "law" of nature—not a law governing allowed changes in the state of the world from one moment to the next following its inception but a law governing the initial conditions themselves.

There are several interesting examples of this sort. One we have already met: the no-boundary condition proposed by James Hartle and Stephen Hawking. As noted, there are rival specifications of the initial state which lead to quite different conclusions—among them the one proposed by Alex Vilenkin, which is shown in figure 6.7. We can also imagine an initial state that seems natural in a different sense: a completely random state. Finally, there is the prescription suggested by Roger Penrose. He proposes that there is a way to measure the level of disorder in the gravitational field of the universe—a universal "gravitational entropy" that increases in accord with the second law of thermodynamics. It seems very likely that such an entropy does exist. Hawking has shown that the gravitational fields of black holes have thermodynamic properties—but black holes are not expanding in time, as our universe is, and we do not yet know what determines the gravitational entropy of an *expanding* universe. For a black hole, the answer is simple: the surface area of the boundary of the black hole determines its gravitational entropy. Penrose and others have suggested that some measure of the regularity of the universe associated with its area might tell us its gravitational entropy. If the expansion rate were the same in every direction and in every place, the entropy would be very small. If the expansion were chaotically different from place to place and from one direction to another, the entropy would be high.

Regardless of the exact indicator of the gravitational entropy, we see that if it increases with time then the initial state of the universe was one of very low, or even zero, gravitational entropy. If we could identify precisely what aspect of the universe tells us its gravitational entropy, we could figure out some of the consequences of its being very low when the universe began. So far, we have not been able to do this.

None of these "principles" concerning the origin of the universe is particularly recommended as the way to solve the greatest problem of cosmology. All are highly speculative. They are ideas for ideas. However, there is an important proviso attached to *any* attempt to explain from first principles the structure of the universe we observe today.

Recall that we have distinguished the universe as a whole from that finite part of it over which light has had time to travel to us since it began. This we have called the

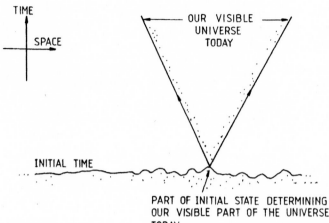

FIGURE 8.2

Our visible universe today expands at the speed of light from a point in the initial state of the universe. The observed portion of the universe is determined by conditions at that point, not by the average conditions of the entire initial state which itself has been dictated by some "principle" governing initial conditions.

· · · · · · · · · ·

"visible universe." The visible universe is necessarily finite in size. When we say that we want to explain the structure of the universe, we mean that we want to explain the form of the visible universe. *The* universe, however, may be finite or infinite in extent. We can never know. If it is infinite in extent, then the visible universe will always be an infinitesimal part of the whole.

These limitations raise a big question mark over the utility of grand principles about the initial state of *the* universe. In our picture of the expansion of the universe, the visible part has expanded from some point or tiny region of the initial state, as shown in figure 8.2.

The structure of the visible universe today is just the expanded image of conditions in some tiny region of the initial state. The grand "principle," on the other hand, gives us an average prescription of the initial state of the entire universe. This prescription may be correct, but it is not what we need to understand the visible universe. We need to know about the particular local state of affairs that existed in the tiny region of the initial state that grew into our visible universe. This region might have been atypical in some way, since it has expanded into a state in which observers can evolve. We have seen that the evolution of observers requires the region to possess many unusual properties. The universe may have begun in a state of minimum gravitational entropy, but this is unlikely to explain the visible universe's structure, because the visible universe may have resulted from the expansion of an anomalous fluctuation—and not from the average state that the minimum-entropy condition prescribed. Moreover, the restriction of our empirical knowledge about the universe to the visible region means that we can never test the consequences of a prescription for the entire initial state of the universe. We see only the evolutionary consequences of a tiny part of that initial state. One day we may be able to say something about the origins of our own cosmic neighborhood. But we can never know the origins of *the* universe. The deepest secrets are the ones that keep themselves.

....

CHAPTER I. THE UNIVERSE IN A NUTSHELL

Barrow, John D., and Joseph Silk, *The Left Hand of Creation: The Origin and Evolution of the Universe*, 2nd ed. (New York: Oxford University Press, 1994).

Cornell, James, ed., *Bubbles, Voids, and Bumps in Time: The New Cosmology* (Cambridge: Cambridge University Press, 1989).

Ferris, Timothy, *Coming of Age in the Milky Way* (New York: William Morrow, 1988).

Gribbin, John, *In Search of the Big Bang* (London: Heinemann, 1986).

Harrison, Edward R., *Cosmology: The Science of the Universe* (Cambridge: Cambridge University Press, 1981).

Long, Charles H., *Alpha: The Myths of Creation* (New York: George Braziller, 1963).

Muller, Richard A., "The Cosmic Background Radiation and the New Aether Drift," *Scientific American*, May 1978, pp. 64–74.

Munitz, Milton K., ed., *Theories of the Universe: From Babylonian Myth to Modern Science* (New York: The Free Press, 1957).

Rowan Robinson, Michael, *Universe* (London: Longman, 1990).

Silk, Joseph, *The Big Bang*, 2nd ed. (San Francisco: W. H. Freeman, 1988).

CHAPTER 2. THE GREAT UNIVERSAL CATALOG

Barrow, John D., and Frank J. Tipler, *The Anthropic Cosmological Principle* (Oxford: Oxford University Press, 1986).

Berendzen, Richard, Richard Hart, and Daniel Seeley, *Man Discovers the Galaxies* (New York: Science History Publications, 1976).

Bertotti, Bruno, Roberto Balbinot, Silvio Bergia, and Andrea Messina, *Modern Cosmology in Retrospect* (Cambridge: Cambridge University Press, 1990).

Brush, Stephen G., *The Kind of Motion We Call Heat*, 2 vols. (Amsterdam: North-Holland, 1976).

North, John D., *The Measure of the Universe* (New York: Dover, 1990).

CHAPTER 3. THE SINGULARITY AND OTHER PROBLEMS

Close, Frank E., *The Cosmic Onion: Quarks and the Nature of the Universe* (London: Heinemann, 1983).

Davies, Paul C. W., *Space and Time in the Modern Universe* (Cambridge: Cambridge University Press, 1977).

———, *The Edge of Infinity* (London: Dent, 1981).

Lederman, Leon, and David N. Schramm, *From Quarks to the Cosmos: Tools of Discovery* (San Francisco: W. H. Freeman, 1989).

Tayler, Roger J., *Hidden Matter* (Chichester: Ellis Horwood, 1991).

Weinberg, Steven, *The First Three Minutes: A Modern View of the Origin of the Universe*, updated ed. (New York: Basic Books, 1988).

Wheeler, John A., *Gravity and Spacetime* (San Francisco: W. H. Freeman, 1990).

CHAPTER 4. INFLATION AND THE PARTICLE PHYSICISTS

Barrow, John D., *The World within the World: 2nd ed.* (Oxford: Oxford University Press, 1994).

Carrigan, Richard A., and W. Peter Trower, *Particle Physics in the Cosmos: Readings from Scientific American* (San Francisco: W. H. Freeman, 1989).

———, *Particles and Forces: At the Heart of the Matter* (San Francisco: W. H. Freeman, 1990).

Georgi, Howard, "Grand Unified Theories," in Davies, Paul C.

W., ed., *The New Physics* (Cambridge: Cambridge University Press, 1989).

Guth, Alan H., and Paul Steinhardt, "The Inflationary Universe," *Scientific American*, May 1984, pp. 116–120.

Krauss, Lawrence M., *The Fifth Essence: The Search for Dark Matter in the Universe* (New York: Basic Books, 1989).

Pagels, Heinz R., *Perfect Symmetry* (London: M. Joseph, 1985).

Trefil, James, *The Moment of Creation* (New York: Scribners, 1983).

Tryon, Edward P., "Cosmic Inflation," *The Encyclopedia of Physical Science and Technology*, vol. 3 (New York: Academic Press, 1987).

Zee, Anthony, *Fearful Symmetry: The Search for Beauty in Modern Physics* (New York: Macmillan, 1986).

CHAPTER 5. INFLATION AND THE COBE SEARCH

Barrow, John D., *Theories of Everything: The Quest for Ultimate Explanation* (Oxford: Oxford University Press, 1991).

Chown, Marcus, *The Afterglow of Creation* (London: Arrow, 1993).

Davies, Paul C. W., *Other Worlds* (London: Dent, 1980).

Gamow, George, *Mr. Tompkins in Paperback* (Cambridge: Cambridge University Press, 1965).

Gribbin, John, and Martin Rees, *Cosmic Coincidences* (New York: Bantam, 1989).

Hey, Anthony, and Patrick Walters, *The Quantum Universe* (Cambridge: Cambridge University Press, 1987).

Linde, Andrei D., "The Universe: Inflation out of Chaos," *New Scientist*, March 1985, pp. 14–16.

Pagels, Heinz R., *The Cosmic Code: Quantum Physics As the Language of Nature* (New York: Simon & Schuster, 1982).

Powell, C. S., "The Golden Age of Cosmology," *Scientific American*, July 1992, pp. 9–12.

Rowan Robinson, Michael, *Ripples in Time* (San Francisco: W. H. Freeman, 1993).

Smoot, George, and Keay Davidson, *Wrinkles in Time* (New York: William Morrow, 1994).

CHAPTER 6. TIME—AN EVEN BRIEFER HISTORY

Grünbaum, Adolf, "The Pseudo-problem of Creation in Cosmology," *Philosophy of Science* 56 (1989): 373.

Hartle, James B., and Stephen W. Hawking, "Wave Function of the Universe," *Physical Review D* 28 (1983): 2960.

Hawking, Stephen W., *A Brief History of Time: From the Big Bang to Black Holes* (New York: Bantam, 1988).

————, "The Edge of Spacetime," in Davies, Paul C. W., ed., *The New Physics* (Cambridge: Cambridge University Press, 1989).

Isham, Christopher J., "Creation of the Universe as a Quantum Process," in Russell, Robert J., William Stoeger, and George V. Coyne, eds., *Physics, Philosophy, and Theology* (Notre Dame, Ind.: University of Notre Dame Press, 1988).

Vilenkin, Alex, "Boundary Conditions in Quantum Cosmology," *Physical Review D* 33 (1982): 3560.

————, "Creation of Universes from Nothing," *Physics Letters B* 117 (1982): 25.

CHAPTER 7. INTO THE LABYRINTH

Barrow, John D., and Frank J. Tipler, *The Anthropic Cosmological Principle* (Oxford: Oxford University Press, 1986).

Blau, Steven K., and Alan H. Guth, "Inflationary Cosmology," in Hawking, Stephen W., and Werner Israel, eds., *300 Years of Gravitation* (Cambridge: Cambridge University Press, 1987).

Coleman, Sidney, "Why There Is Something Rather Than Nothing: A Theory of the Cosmological Constant," *Nuclear Physics B* 310 (1988): 643.

Drees, Willem B., *Beyond the Big Bang: Quantum Cosmology and God* (La Salle, Ill.: Open Court, 1990).

Halliwell, Jonathan J., "Quantum Cosmology and the Creation of the Universe," *Scientific American*, December 1991, pp. 28–35.

Hawking, Stephen W., "Wormholes on Spacetime," *Physical Review D* 37 (1988): 904.

Hoyle, Fred, *Galaxies, Nuclei, and Quasars* (London: Heinemann, 1964).

Leslie, John, *Universes* (London: Macmillan, 1989).

Weinberg, Steven, "The Cosmological Constant Problem," *Reviews of Modern Physics* 61 (1989): 1.

CHAPTER 8. NEW DIMENSIONS

Barrow, John D., "Observational Limits on the Time-evolution of Extra Spatial Dimensions," *Physical Review D* 35 (1987): 1805.

——, *Theories of Everything: The Quest for Ultimate Explanation* (Oxford: Oxford University Press, 1991).

——, "Unprincipled Cosmology," *Quarterly Journal of the Royal Astronomical Society* 34 (1993): 117.

Davies, Paul C. W., and Julian R. Brown, *Superstrings: A Theory of Everything?* (Cambridge: Cambridge University Press, 1988).

Green, Michael B., "Superstrings," *Scientific American*, September 1986, pp. 48–60.

Peat, F. David, *Superstrings and the Search for a Theory of Everything* (Chicago: Contemporary Books, 1988).

Penrose, Roger, *The Emperor's New Mind: Concerning Computers, Minds, and the Laws of Physics* (Oxford: Oxford University Press, 1989).

Printed in the United States
34901LVS00001B/19

9 780465 053148